HSC Year 12

MATHEMATICS EXTENSION 1

KAREN MAN | ASHLEIGH DELLA MARTA

SERIES EDITOR: ROBERT YEN

2020 UPDATED SYLLABUS • 2020 UPDATED SYLLABUS •

A+

+ topic summaries
+ graded practice questions with worked solutions
+ HSC exam topic grids (2011–2020)

STUDY NOTES

A+ HSC Mathematics Extension 1 Study Notes
1st Edition
Karen Man
Ashleigh Della Marta
ISBN 9780170459242

Publishers: Robert Yen, Kirstie Irwin
Project editor: Tanya Smith
Cover design: Nikita Bansal
Text design: Alba Design
Project designer: Nikita Bansal
Permissions researcher: Corrina Gilbert
Production controller: Karen Young
Typeset by: Nikki M Group Pty Ltd

ISBN 978 0 17 045924 2

Cengage Learning Australia
Level 7, 80 Dorcas Street
South Melbourne, Victoria Australia 3205

Cengage Learning New Zealand
Unit 4B Rosedale Office Park
331 Rosedale Road, Albany, North Shore 0632, NZ

For learning solutions, visit **cengage.com.au**

Printed in China by 1010 Printing International Limited.
1 2 3 4 5 6 7 25 24 23 22 21

ABOUT THIS BOOK

Introducing *A+ HSC Year 12 Mathematics*, a new series of study guides designed to help students revise the topics of the new HSC maths courses and achieve success in their exams. *A+* is published by Cengage, the educational publisher of *Maths in Focus* and *New Century Maths*.

For each HSC maths course, Cengage has developed a STUDY NOTES book and a PRACTICE EXAMS book. These study guides have been written by experienced teachers who have taught the new courses, some of whom are involved in HSC exam marking and writing. This is the first study guide series to be published after the first HSC exams of the new courses in 2020, so it incorporates the latest changes to the syllabus and exam format.

This book, *A+ HSC Year 12 Mathematics Extension 1 Study Notes,* contains topic summaries and graded practice questions, grouped into 6 broad topics, addressing the outcomes in the Mathematics Extension 1 syllabus. The topic-based structure means that this book can be used for revision after a topic has been covered in the classroom, as well as for course review and preparation for the trial and HSC exams. Each topic chapter includes a review of the main mathematical concepts, and multiple-choice and short-answer questions with worked solutions. Past HSC examination questions have been included to provide students with the opportunity to see how they will be expected to show their mathematical understanding in the exams. An HSC exam topic grid (2011–2020) guides students to where and how each topic has been tested in past HSC exams.

Mathematics Extension 1 Year 12 topics

1. Mathematical induction
2. Vectors
3. Trigonometric equations
4. Further integration
5. Differential equations
6. The binomial distribution

This book contains:

- Concept map (see p. 20 for an example)
- Glossary and digital flashcards (see p. 21 for an example)
- Topic summary, addressing key outcomes of the syllabus (see p. 22 for an example)
- Practice set 1: 20 multiple-choice questions (see p. 31 for an example)
- Practice set 2: 20 short-answer questions (see p. 35 for an example)
- Questions graded by level of difficulty: foundation ●○○, moderate ●●○, complex ●●●
- Worked solutions to both practice sets
- HSC exam topic grid (2011–2020) (see p. 51 for an example)

The companion A+ PRACTICE EXAMS book contains topic exams and practice HSC exam papers, both of which are written and formatted in the style of the HSC exam paper, with spaces for students to write answers. Worked solutions are provided, along with the authors' expert comments and advice, including how each exam question is marked. As a special bonus, the worked solutions to the 2020 HSC exam paper have been included.

This A+ STUDY NOTES book will become a staple resource in your study in the lead-up to your final HSC exams. Revisit it throughout Year 12 to ensure that you do not forget key concepts and skills. Good luck!

CONTENTS

ABOUT THIS BOOK III
YEAR 12 COURSE OVERVIEW VI
SYLLABUS REFERENCE GRID VIII
ABOUT THE AUTHORS VIII
A+ DIGITAL FLASHCARDS VIII
HSC EXAM FORMAT IX
STUDY AND EXAM ADVICE X
MATHEMATICAL VERBS XIV
SYMBOLS AND ABBREVIATIONS XV

CHAPTER 1

MATHEMATICAL INDUCTION

Concept map	2
Glossary	3
Topic summary	4
Practice sets tracking grid	6
Practice set 1: Multiple-choice questions	7
Practice set 2: Short-answer questions	9
Practice set 1: Worked solutions	11
Practice set 2: Worked solutions	12
HSC exam topic grid (2011–2020)	18

VECTORS

Concept map	20
Glossary	21
Topic summary	22
Practice sets tracking grid	30
Practice set 1: Multiple-choice questions	31
Practice set 2: Short-answer questions	35
Practice set 1: Worked solutions	40
Practice set 2: Worked solutions	43
HSC exam topic grid (2011–2020)	51

TRIGONOMETRIC EQUATIONS

Concept map	53
Glossary	54
Topic summary	55
Practice sets tracking grid	57
Practice set 1: Multiple-choice questions	58
Practice set 2: Short-answer questions	61

Practice set 1: Worked solutions 63
Practice set 2: Worked solutions 68
HSC exam topic grid (2011–2020) 77

CHAPTER 4 FURTHER INTEGRATION
Concept map 79
Glossary 80
Topic summary 81
Practice sets tracking grid 86
Practice set 1: Multiple-choice questions 87
Practice set 2: Short-answer questions 90
Practice set 1: Worked solutions 94
Practice set 2: Worked solutions 98
HSC exam topic grid (2011–2020) 106

CHAPTER 5 DIFFERENTIAL EQUATIONS
Concept map 108
Glossary 109
Topic summary 110
Practice sets tracking grid 113
Practice set 1: Multiple-choice questions 114
Practice set 2: Short-answer questions 120
Practice set 1: Worked solutions 125
Practice set 2: Worked solutions 128
HSC exam topic grid (2011–2020) 136

CHAPTER 6 THE BINOMIAL DISTRIBUTION
Concept map 138
Glossary 139
Topic summary 140
Practice sets tracking grid 143
Practice set 1: Multiple-choice questions 144
Practice set 2: Short-answer questions 147
Practice set 1: Worked solutions 152
Practice set 2: Worked solutions 156
HSC exam topic grid (2011–2020) 162

HSC EXAM REFERENCE SHEET 163
INDEX 167

YEAR 12 COURSE OVERVIEW

See each concept map printed in full size at the beginning of each chapter.

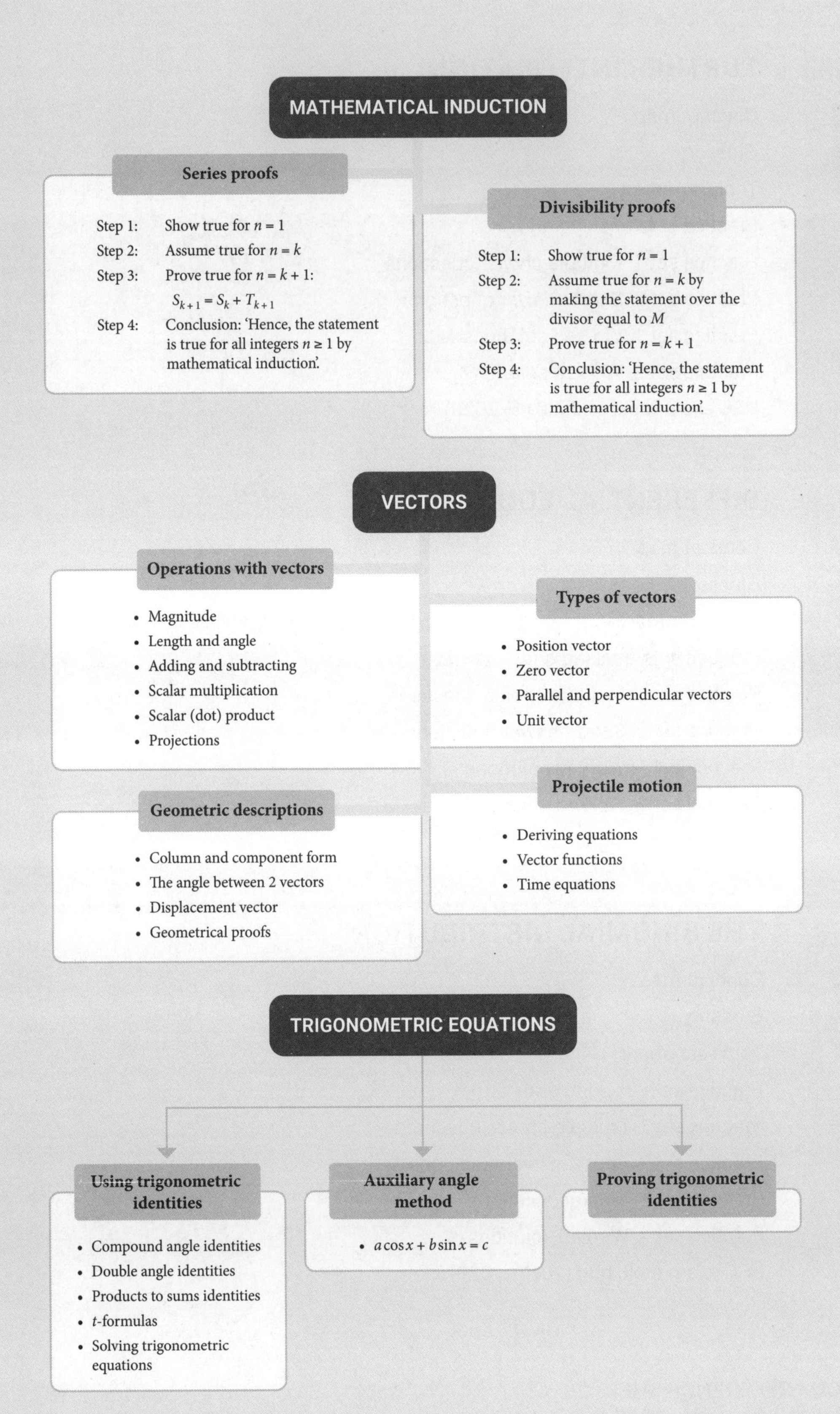

FURTHER INTEGRATION

Integration by substitution

- Let $u = \ldots$
- Given the substitution
- Choosing the substitution
- Trigonometric substitutions

Trigonometric integrals

- Integrating $\sin^2 x$ and $\cos^2 x$
- Trigonometric identities
- Products to sums formulas

Inverse trigonometric functions

- Differentiating inverse functions
- Differentiating inverse trigonometric functions
- Integrals involving inverse functions

Areas of integration

- Areas about the x-axis
- Areas about the y-axis

Volumes of solids of revolution

- Volumes about the x-axis
- Volumes about the y-axis

DIFFERENTIAL EQUATIONS

Solving differential equations

$\frac{dy}{dx} = f(x)$

$\frac{dy}{dx} = g(y)$

$\frac{dy}{dx} = f(x)\,g(y)$: separation of variables

Direction fields

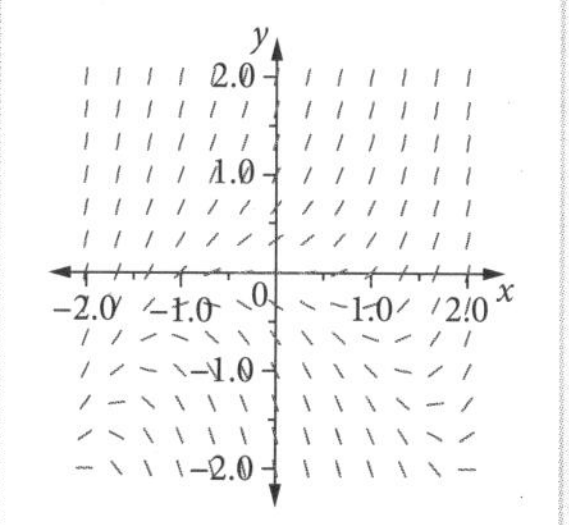

Application of differential equations

- Exponential growth and decay
- Newton's law of cooling
- Logistic equations

THE BINOMIAL DISTRIBUTION

Binomial distribution

- Bernoulli distributions
- Binomial distributions
- Mean and variance

$E(X) = np$

$\text{Var}(X) = np(1 - p)$

Binomial probability

$P(X = r) = {}^{n}C_r p^r (1 - p)^{n - r}$

$X \sim \text{Bin}(n, p)$

Sample proportions

- Sample proportion distributions

$E(\hat{p}) = p$

$\text{Var}(\hat{p}) = \frac{pq}{n}$

- Normal approximation

SYLLABUS REFERENCE GRID

Topic and subtopics	***A+ HSC Year 12 Mathematics Extension 1 Study Notes* chapter**
PROOF	
ME-P1 Proof by mathematical induction	1 Mathematical induction
VECTORS	
ME-V1 Introduction to vectors V1.1 Introduction to vectors V1.2 Further operations with vectors V1.3 Projectile motion	2 Vectors
TRIGONOMETRIC FUNCTIONS	
ME-T3 Trigonometric equations	3 Trigonometric equations
CALCULUS	
ME-C2 Further calculus skills	4 Further integration
ME-C3 Applications of calculus C3.1 Further area and volumes of solids of revolution C3.2 Differential equations	 4 Further integration 5 Differential equations
STATISTICAL ANALYSIS	
ME-S1 The binomial distribution S1.1 Bernoulli and binomial distributions S1.2 Normal approximation for the sample proportion	6 The binomial distribution

ABOUT THE AUTHORS

Karen Man teaches at Our Lady of the Sacred Heart College, Kensington. She has presented HSC workshops for students in Sydney Catholic schools and works on the team that writes and edits MANSW's annual HSC mathematics exam solutions.

Ashleigh Della Marta is assistant Head of Mathematics at Waverley College.

A+ DIGITAL FLASHCARDS

Revise key terms and concepts online with the A+ Flashcards. Each topic glossary in this book has a corresponding deck of digital flashcards you can use to test your understanding and recall. Just scan the QR code or type the URL into your browser to access them.

Note: You will need to create a free *NelsonNet* account.

HSC EXAM FORMAT

Mathematics Extension 1 students complete two HSC exams: **Mathematics Advanced** and **Mathematics Extension 1**.

The following information about the exams was correct at the time of printing in 2021. Please check the NESA website in case it has changed. Visit www.educationstandards.nsw.edu.au, select 'Year 11–Year 12', 'Syllabuses A–Z', 'Mathematics Advanced/Extension 1', then 'Assessment and Reporting'. Scroll down to 'HSC examination specifications'.

Mathematics Advanced HSC exam

	Questions	Marks	Recommended time
Section I	10 multiple-choice questions Mark answers on the multiple-choice answer sheet.	10	15 min
Section II	Approx. 24 short-answer questions, including 2 or more questions worth 4 or 5 marks. Write answers on the lines provided on the paper.	90	2 h 45 min
Total		100	3 h

- Reading time: 10 minutes; use this time to preview the whole exam.
- Working time: 3 hours
- Questions focus on Year 12 outcomes but Year 11 knowledge may be examined.
- Answers are to be written on the question paper.
- A reference sheet is provided at the back of the exam paper, and also this book, containing common formulas.
- Common questions with the Mathematics Standard 2 HSC exam: 20–25 marks
- The 4- or 5-mark questions are usually complex problems that require many steps of working and careful planning.
- To help you plan your time, the halfway point of Section II is marked by a notice at the bottom of the relevant page; for example, 'Questions 11–23 are worth 46 marks in total'.
- Having 3 hours for a total of 100 marks means that you have an average of 1.8 minutes per mark (or approximately 5 minutes for 3 marks).
- If you budget 15 minutes for Section I, and 1 hour 15 minutes for each half of Section II, you will have 15 minutes at the end to check over your work and/or complete questions you missed.

Mathematics Extension 1 HSC exam

	Questions	Marks	Recommended time
Section I	10 multiple-choice questions	10	15 min
Section II	4 multi-part short-answer questions, average 15 marks each, including questions worth 4 or 5 marks	60	1 h 45 min
Total		70	2 h

- Reading time: 10 minutes; Working time: 2 hours
- Answers are to be written in separate answer booklets.
- Having 2 hours for a total of 70 marks means that you have an average of 1.7 minutes per mark (or approximately 5 minutes for 3 marks).
- If you budget 15 minutes for Section I, then 20 minutes per question for Section II, you will have 25 minutes at the end to check over your work and/or complete questions you missed.

STUDY AND EXAM ADVICE

A journey of a thousand miles begins with a single step.

Lao Tzu (c. 570–490 BCE), Chinese philosopher

I've always believed that if you put in the work, the results will come.

Michael Jordan (1963–), American basketball player

Four PRACtical steps for maths study

1. Practise your maths

- Do your homework.
- Learning maths is about mastering a collection of skills.
- You become successful at maths by doing it more, through regular practice and learning.
- Aim to achieve a high level of understanding.

2. Rewrite your maths

- Homework and study are not the same thing. Study is your private 'revision' work for strengthening your understanding of a subject.
- Before you begin any questions, make sure you have a thorough understanding of the topic.
- Take ownership of your maths. Rewrite the theory and examples in your own words.
- Summarise each topic to see the 'whole picture' and know it all.

3. Attack your maths

- All maths knowledge is interconnected. If you don't understand one topic fully, then you may have trouble learning another topic.
- Mathematics is not an HSC course you can learn 'by halves' – you have to know it all!
- Fill in any gaps in your mathematical knowledge to see the 'whole picture'.
- Identify your areas of weakness and work on them.
- Spend most of your study time on the topics you find difficult.

4. Check your maths

- After you have mastered a maths skill, such as graphing a quadratic equation, no further learning or reading is needed, just more practice.
- Compared to other subjects, the types of questions asked in maths exams are conventional and predictable.
- Test your understanding with revision exercises, practice papers and past exam papers.
- Develop your exam technique and problem-solving skills.
- Go back to steps 1–3 to improve your study habits.

Topic summaries and concept maps

Summarise each topic when you have completed it, to create useful study notes for revising the course, especially before exams. Use a notebook or folder to list the important ideas, formulas, terminology and skills for each topic. This book is a good study guide, but educational research shows that effective learning takes place when you rewrite learned knowledge in your own words.

A good topic summary runs for 2 to 4 pages. It is a condensed, personalised version of your course notes. This is your interpretation of a topic, so include your own comments, symbols, diagrams, observations and reminders. Highlight important facts using boxes and include a glossary of key words and phrases.

A concept map or mind map is a topic summary in graphic form, with boxes, branches and arrows showing the connections between the main ideas of the topic. This book contains examples of concept maps. The topic name is central to the map, with key concepts or subheadings listing important details and formulas. Concept maps are powerful because they present an overview of a topic on one large sheet of paper. Visual learners absorb and recall information better when they use concept maps.

When compiling a topic summary, use your class notes, your textbook and this study guide. Ask your teacher for a copy of the course syllabus or the school's teaching program, which includes the objectives and outcomes of every topic in dot point form.

Attacking your weak areas

Most of your study time should be spent on attacking your weak areas to fill in any gaps in your maths knowledge. Don't spend too much time on work you already know well, unless you need a confidence boost! Ask your teacher, use this book or your textbook to improve the understanding of your weak areas and to practise maths skills. Use your topic summaries for general revision, but spend longer study periods on overcoming any difficulties in your mastery of the course.

Practising with past exam papers

Why is practising with past exam papers such an effective study technique? It allows you to become familiar with the format, style and level of difficulty expected in an HSC exam, as well as the common topic areas tested. Knowing what to expect helps alleviate exam anxiety. Remember, mathematics is a subject in which the exam questions are fairly predictable. The exam writers are not going to ask too many unusual questions. By the time you have worked through many past exam papers, this year's HSC exams won't seem that much different.

Don't throw your old exam papers away. Use them to identify your mistakes and weak areas for further study. Revising topics and then working on mixed questions is a great way to study maths. You might like to complete a past HSC exam paper under timed conditions to improve your exam technique.

Past HSC exam papers are available at the NESA website: visit www.educationstandards.nsw.edu.au and select 'Year 11 – Year 12', 'HSC exam papers'. NESA marking feedback and guidelines can also be viewed there. Cengage has also published *A+ HSC Year 12 Mathematics Extension 1 Practice Exams*, containing topic exams and practice HSC exam papers. You can find past HSC exam papers with solutions online, in bookstores, at the Mathematical Association of NSW (www.mansw.nsw.edu.au) and at your school (ask your teacher) or library.

Preparing for an exam

- Make a study plan early; don't leave it until the last minute.
- Read and revise your topic summaries.
- Work on your weak areas and learn from your mistakes.
- Don't spend too much time studying concepts you know already.
- Revise by completing revision exercises and past exam papers or assignments.
- Vary the way you study so that you don't become bored: ask someone to quiz you, voice-record your summary, design a poster or concept map, or explain the concept to someone.
- Anticipate the exam:
  - How many questions will there be?
  - What are the types of questions: multiple-choice, short-answer, long-answer, problem-solving?
  - Which topics will be tested?
  - How many marks are there in each section?
  - How long is the exam?
  - How much time should I spend on each question/section?
  - Which formulas are on the reference sheet and how do I use them in the exam?

During an exam

1. Bring all of your equipment, including a ruler and calculator (check that your calculator works and is in RADIANS mode for trigonometric functions and DEGREES for trigonometric measurements). A highlighter pen may help for tables, graphs and diagrams.
2. Don't worry if you feel nervous before an exam – this is normal and can help you to perform better; however, being too casual or too anxious can harm your performance. Just before the exam begins, take deep, slow breaths to reduce any stress.
3. Write clearly and neatly in black or blue pen, not red. Use a pencil only for diagrams and constructions.
4. Use the **reading time** to browse through the exam to see the work that is ahead of you and the marks allocated to each question. Doing this will ensure you won't miss any questions or pages. Note the harder questions and allow more time for working on them. Leave them if you get stuck, and come back to them later.
5. Attempt every question. It is better to do most of every question and score some marks, rather than ignore questions completely and score 0 for them. Don't leave multiple-choice questions unanswered! Even if you guess, you have a chance of being correct.
6. Easier questions are usually at the beginning, with harder ones at the end. Do an easy question first to boost your confidence. Some students like to leave multiple-choice questions until last so that, if they run out of time, they can make quick guesses. However, some multiple-choice questions can be quite difficult.
7. Read each question and identify what needs to be found and what topic/skill it is testing. The number of marks indicates how much time and working out is required. Highlight any important keywords or clues. Do you need to use the answer to the previous part of the question?
8. After reading each question, and before you start writing, spend a few moments planning and thinking.
9. You don't need to be writing all of the time. What you are writing may be wrong and a waste of time. Spend some time considering the best approach.
10. Make sure each answer seems reasonable and realistic, especially if it involves money or measurement.

11. Show all necessary working, write clearly, draw big diagrams, and set your working out neatly. Write solutions to each part underneath the previous step so that your working out goes down the page, not across.
12. Use a ruler to draw (or read) half-page graphs with labels and axes marked, or to measure scale diagrams.
13. Don't spend too much time on one question. Keep an eye on the time.
14. Make sure you have answered the question. Did you remember to round the answer and/or include units? Did you use all of the relevant information given?
15. If a hard question is taking too long, don't get bogged down. If you're getting nowhere, retrace your steps, start again, or skip the question (circle it) and return to it later with a clearer mind.
16. If you make a mistake, cross it out with a neat line. Don't scribble over it completely or use correction fluid or tape (which is time-consuming and messy). You may still score marks for crossed-out work if it is correct, but don't leave multiple answers! Keep track of your answer booklets and ask for more writing paper if needed.
17. Don't cross out or change an answer too quickly. Research shows that often your first answer is the correct one.
18. Don't round your answer in the middle of a calculation. Round at the end only.
19. Be prepared to write words and sentences in your answers, but don't use abbreviations that you've just made up. Use correct terminology and write one or two sentences for 2 or 3 marks, not mini-essays.
20. If you have time at the end of the exam, double-check your answers, especially for the more difficult questions or questions you are uncertain about.

Ten exam habits of the best HSC students

1. Has clear and careful working and checks their answers
2. Has a strong understanding of basic algebra and calculation
3. Reads (and answers) the whole question
4. Chooses the simplest and quickest method
5. Checks that their answer makes sense or sounds reasonable
6. Draws big, clear diagrams with details and labels
7. Uses a ruler for drawing, measuring and reading graphs
8. Can explain answers in words when needed, in one or two clear sentences
9. Uses the previous parts of a question to solve the next part of the question
10. Rounds answers at the end, not before

Further resources

Visit the NESA website (www.educationstandards.nsw.edu.au) for the following resources. Select 'Year 11 – Year 12' and then 'Syllabuses A–Z' or 'HSC exam papers'.

- Mathematics Advanced and Extension 1 syllabuses
- Past HSC exam papers, including marking feedback and guidelines
- Sample HSC questions/exam papers and marking guidelines

Before 2020, 'Mathematics Advanced' was called 'Mathematics' and although 'Mathematics Extension 1' had the same name, it was a different course with some topics that no longer exist. For these exam papers, select 'Year 11 – Year 12', 'Resources archive', 'HSC exam papers archive'.

MATHEMATICAL VERBS

A glossary of 'doing words' common in maths problems and HSC exams

analyse
study in detail the parts of a situation

apply
use knowledge or a procedure in a given situation

calculate
See **evaluate**

classify/identify
state the type, name or feature of an item or situation

comment
express an observation or opinion about a result

compare
show how two or more things are similar or different

complete
fill in detail to make a statement, diagram or table correct or finished

construct
draw an accurate diagram

convert
change from one form to another, for example, from a fraction to a decimal, or from kilograms to grams

decrease
make smaller

describe
state the features of a situation

estimate
make an educated guess for a number, measurement or solution, to find roughly or approximately

evaluate/calculate
find the value of a numerical expression, for example, 3×8^2 or $4x + 1$ when $x = 5$

expand
remove brackets in an algebraic expression, for example, expanding $3(2y + 1)$ gives $6y + 3$

explain
describe why or how

give reasons
show the rules or thinking used when solving a problem. *See also* **justify**

graph
display on a number line, number plane or statistical graph

hence find/prove
calculate an answer or demonstrate a result using previous answers or information supplied

identify
See **classify**

increase
make larger

interpret
find meaning in a mathematical result

justify
give reasons or evidence to support your argument or conclusion. *See also* **give reasons**

measure
determine the size of something, for example, using a ruler to determine the length of a pen

prove
See **show/prove that**

recall
remember and state

show/prove that
(in questions where the answer is given) use calculation, procedure or reasoning to demonstrate that an answer or result is true

simplify
express a result such as a ratio or algebraic expression in its most basic, shortest, neatest form

sketch
draw a rough diagram that shows the general shape or ideas (less accurate than **construct**)

solve
calculate the value(s) of an unknown pronumeral in an equation or inequality

state
See **write**

substitute
replace part of an expression with another, equivalent expression

verify
check that a solution or result is correct, usually by substituting back into an equation or referring back to the problem

write/state
give an answer, formula or result without showing any working or explanation (This usually means that the answer can be found mentally, or in one step)

SYMBOLS AND ABBREVIATIONS

Symbol	Meaning
$=$	is equal to
$\neq$	is not equal to
$\approx$	is approximately equal to
$<$	is less than
$>$	is greater than
$\leq$	is less than or equal to
$\geq$	is greater than or equal to
$(\ )$	parentheses, round brackets
$[\ ]$	(square) brackets
$\{\ \}$	braces
$\pm$	plus or minus
π	pi = 3.14159 …
$\equiv$	is congruent/identical to
$^\circ$	degree
$\angle$	angle
Δ	triangle, the discriminant
$\parallel$	is parallel to
$\perp$	is perpendicular to
x^2	x squared, $x \times x$
x^3	x cubed, $x \times x \times x$
$\cup$	union
$\cap$	intersection
∞	infinity
$\lvert x \rvert$	absolute value or magnitude of x
$\underset{\sim}{v}$	the vector v
$\overrightarrow{AB}$	the vector AB
$\underset{\sim}{u} \cdot \underset{\sim}{v}$	the scalar product of $\underset{\sim}{u}$ and $\underset{\sim}{v}$
$\text{proj}_{\underset{\sim}{u}} \underset{\sim}{v}$	the projection of $\underset{\sim}{v}$ onto $\underset{\sim}{u}$
$\lim\limits_{h \to 0}$	the limit as $h \to 0$
$\frac{dy}{dx}, y', f'(x)$	the first derivative of $y, f(x)$
$\frac{d^2y}{dx^2}, y'', f''(x)$	the second derivative of $y, f(x)$
$\int f(x)\,dx$	the integral of $f(x)$
$f^{-1}(x)$	the inverse function of $f(x)$
$\sin^{-1}$, arcsin	the inverse sine function
Σ	sigma, the sum of
$\therefore$	therefore

Symbol	Meaning
$[a, b], a \leq x \leq b$	the interval of x-values from a to b (including a and b)
$(a, b), a < x < b$	the interval of x-values between a and b (excluding a and b)
$P(E)$	the probability of event E occurring
$P(\overline{E})$	the probability of event E not occurring
$A \cup B$	A union B, A or B
$A \cap B$	A intersection B, A and B
$P(A \mid B)$	the probability of A given B
$n!$	n factorial, $n(n-1)(n-2) \ldots \times 1$
${}^nC_r, \binom{n}{r}$	the number of combinations of r objects from n objects
${}^nP_r$	the number of permutations of r objects from n objects
PDF	probability density function
CDF	cumulative distribution function
$X \sim \text{Bin}(n, p)$	X is a random variable of the binomial distribution
$\hat{p}$	sample proportion
LHS	left-hand side
RHS	right-hand side
p.a.	per annum (per year)
cos	cosine ratio
sin	sine ratio
tan	tangent ratio
$\overline{x}$	the mean
$\mu = E(X)$	the population mean, expected value
σ	the standard deviation
$\text{Var}(X) = \sigma^2$	the variance
Q_1	first quartile or lower quartile
Q_2	median (second quartile)
Q_3	third quartile or upper quartile
IQR	interquartile range
α	alpha
θ	theta
m	gradient
RTP	required to prove

9780170459242

A+ HSC YEAR 12 MATHEMATICS

STUDY NOTES

Authors:

Tania Eastcott
Rachel Eastcott

Sarah Hamper

Karen Man
Ashleigh Della Marta

Jim Green
Janet Hunter

PRACTICE EXAMS

Authors:

Adrian Kruse

Simon Meli

John Drake

Jim Green
Janet Hunter

CHAPTER 1
MATHEMATICAL INDUCTION

ME-P1 Proof by mathematical induction 4

1

MATHEMATICAL INDUCTION

Series proofs

Step 1: Show true for $n = 1$

Step 2: Assume true for $n = k$

Step 3: Prove true for $n = k + 1$:

$S_{k+1} = S_k + T_{k+1}$

Step 4: Conclusion: 'Hence, the statement is true for all integers $n \geq 1$ by mathematical induction'.

Divisibility proofs

Step 1: Show true for $n = 1$

Step 2: Assume true for $n = k$ by making the statement over the divisor equal to M

Step 3: Prove true for $n = k + 1$

Step 4: Conclusion: 'Hence, the statement is true for all integers $n \geq 1$ by mathematical induction'.

Glossary

divisibility
The feature of a whole number that allows it to be divisible by another whole number with no remainder.

mathematical induction
One method of proving that a statement or theorem is true by using algebra to generalise from a specific case such as $n = 1$. Also called inductive proof or proof by induction.

RTP
'Required to prove', an abbreviation often used at the start of a proof.

statement
A sentence that is true or false but not both.

https://get.ga/a-hsc-maths-ext-1

A+ DIGITAL FLASHCARDS
Revise this topic's key terms and concepts by scanning the QR code or typing the URL into your browser.

Topic summary

Proof by mathematical induction (ME-P1)

- **Mathematical induction** is one method of proving a **statement** (theory) is true for positive integers, n.
- Mathematical induction is a method for proving the formula for a **series**, or for **divisibility** when the aim is to show there is no remainder when a number is divided by another number (divisor).
- Mathematical induction is only appropriate when the mathematical statement involves integers.
- A false proof of induction occurs when any of the steps are false; hence, the statement will not have been proved.

Series proofs: Steps

Step 1: Show that the statement is true for $n = 1$ (or the smallest value of n given).

Step 2: Assume that the statement is true for $n = k$.

Step 3: Using the assumption, prove that the statement is true for the next integer, $n = k + 1$.

$$S_{k+1} = S_k + T_{k+1}$$

Step 4: Conclusion: 'Hence, the statement is true for all integers $n \geq 1$ by mathematical induction'.

Hint
S_{k+1} = sum of $(k + 1)$ terms
S_k = sum of k terms
T_{k+1} = the $(k + 1)$th term

Example 1

Prove by mathematical induction that

$$\frac{1}{1\times 2}+\frac{1}{2\times 3}+\frac{1}{3\times 4}+\cdots+\frac{1}{n(n+1)}=\frac{n}{n+1}$$

for all positive integers n.

Solution

Step 1: Prove the statement is true for $n = 1$ $(T_1 = S_1)$.

$$\begin{aligned} T_1 &= \frac{1}{1\times 2} \\ &= \frac{1}{2} \\ S_1 &= \frac{1}{1+1} \\ &= \frac{1}{2} \\ &= T_1 \end{aligned}$$

So the statement is true for $n = 1$.

Step 2: Assume the statement is true for $n = k$ and then use arrangement in Step 3.

$$\frac{1}{1\times 2}+\frac{1}{2\times 3}+\frac{1}{3\times 4}+\cdots+\frac{1}{k(k+1)}=\frac{k}{k+1} \quad [*]$$

Step 3: Prove the statement is true for $n = k + 1$.

$$\text{RTP } S_{k+1} = \frac{k+1}{(k+1)+1} = \frac{k+1}{k+2}$$

Hint
RTP = 'Required to prove:'

$$\begin{aligned} S_{k+1} &= S_k + T_{k+1} \\ &= \frac{k}{k+1} + \frac{1}{(k+1)(k+2)} \quad \text{from [*]} \\ &= \frac{k(k+2)+1}{(k+1)(k+2)} \\ &= \frac{k^2+2k+1}{(k+1)(k+2)} \\ &= \frac{(k+1)^2}{(k+1)(k+2)} \\ &= \frac{k+1}{k+2}, \text{ as required} \end{aligned}$$

So the statement is true for $n = k + 1$.

Step 4: Conclusion

So the statement is true for all positive integers by mathematical induction.

Divisibility proofs: Steps

Step 1: Show that the statement is true for $n = 1$ (or the smallest value of n given) by showing that the divisor is a factor.

Step 2: Assume that the statement is true for $n = k$ by making the quotient when the statement is divided by the divisor equal to M (where M is an integer).

Step 3: Using the assumption, prove that the statement is true for the next integer, $n = k + 1$, by showing that the divisor is a factor.

Step 4: Conclusion: 'Hence, the statement is true for all integers $n \geq 1$ by mathematical induction'.

Example 2 ©NESA 2017 HSC EXAM, QUESTION 14(a)

Prove by mathematical induction that $8^{2n+1} + 6^{2n-1}$ is divisible by 7, for any integer $n \geq 1$.

Solution

Step 1: Prove the statement is true for $n = 1$.

$$\begin{aligned} 8^{2n+1} + 6^{2n-1} &= 8^3 + 6^1 \qquad (n = 1) \\ &= 512 + 6 \\ &= 518 \\ &= 7 \times 74 \qquad \text{(divisible by 7)} \end{aligned}$$

So the statement is true for $n = 1$.

Step 2: Assume the statement is true for $n = k$ (divisible by 7) and then use arrangement in Step 3.

$$\frac{8^{2k+1} + 6^{2k-1}}{7} = M \quad \text{for some integer } M$$

$$8^{2k+1} + 6^{2k-1} = 7M$$

$$8^{2k+1} = 7M - 6^{2k-1} \qquad [*]$$

Step 3: Prove the statement is true for $n = k + 1$.

Substitute $n = k + 1$:

$$\begin{aligned}
8^{2(k+1)+1} + 6^{2(k+1)-1} &= 8^{2k+3} + 6^{2k+1} \\
&= 8^{2k+1+2} + 6^{2k+1} \\
&= 8^{2k+1}(8^2) + 6^{2k+1} \\
&= (7M - 6^{2k-1})(64) + 6^{2k+1} \quad \text{from [*]} \\
&= 448M - 6^{2k-1}(64) + 6^{2k-1+2} \\
&= 448M - 6^{2k-1}(64) + 6^{2k-1}\, 6^2 \\
&= 448M - 6^{2k-1}(64) + 6^{2k-1}\, (36) \\
&= 448M - 6^{2k-1}(28) \\
&= 7[64M - 6^{2k-1}(4)]
\end{aligned}$$

This is divisible by 7 since M and k are integers.

So the statement is true for $n = k + 1$.

Step 4: Conclusion

So the statement is true for all integers $n \geq 1$ by mathematical induction.

Practice sets tracking grid

Maths is all about repetition, meaning do, do and do again! Each question in the following practice sets, especially the struggle questions (different for everybody!), should be completed at least 3 times correctly. Below is a tracking grid to record your question attempts: ✓ if you answered correctly, ✗ if you didn't.

PRACTICE SET 1: Multiple-choice questions

Question	1st attempt	2nd attempt	3rd attempt	4th attempt	5th attempt
1					
2					
3					

PRACTICE SET 2: Short-answer questions

Question	1st attempt	2nd attempt	3rd attempt	4th attempt	5th attempt
1					
2					
3					
4					
5					
6					
7					
8					
9					
10					
11					
12					
13					
14					
15					
16					
17					
18					

9780170459242

Practice set 1

Multiple-choice questions

Solutions start on page 11.

Question 1 ©NESA 1988 HSC EXAM, QUESTION 3(b) MODIFIED FOR MULTIPLE CHOICE

Mathematical induction is used to prove

$$1^2 + 3^2 + 5^2 + \cdots + (2n-1)^2 = \frac{1}{3}n(2n+1)(2n-1)$$

for all positive integers $n \geq 1$.

Which of the following has an incorrect expression for part of the induction proof?

A Step 1: To prove the statement is true for $n = 1$.

LHS: $1^2 = 1$

$$\begin{aligned} \text{RHS} &= \frac{1}{3} \times 1 \times (2 \times 1 + 1)(2 \times 1 - 1) \\ &= \frac{1}{3} \times 3 \times 1 \\ &= 1 \end{aligned}$$

Result is true for $n = 1$.

B Step 2: Assume the result is true for $n = k$.

$$1^2 + 3^2 + 5^2 + \cdots + (2k-1)^2 = \frac{1}{3}(k+1)(2k+1)(2k-1)$$

C Step 3: To prove the result is true for $n = k + 1$.

$$\begin{aligned} 1^2 + 3^2 + 5^2 + \cdots + (2k-1)^2 + \left(2(k+1)-1\right)^2 &= \frac{1}{3}(k+1)\left(2(k+1)+1\right)\left(2(k+1)-1\right) \\ &= \frac{1}{3}(k+1)(2k+3)(2k+1) \end{aligned}$$

D
$$\begin{aligned} \text{LHS} &= 1^2 + 3^2 + 5^2 + \cdots + (2k-1)^2 + \left(2(k+1)-1\right)^2 \\ &= \frac{1}{3}k(2k+1)(2k-1) + (2k+1)^2 \\ &= \frac{1}{3}(2k+1)\left(k(2k-1) + 3(2k+1)\right) \\ &= \frac{1}{3}(2k+1)(2k^2 - k + 6k + 3) \\ &= \frac{1}{3}(2k+1)(2k^2 + 5k + 3) \\ &= \frac{1}{3}(2k+1)(k+1)(2k+3) \\ &= \text{RHS} \end{aligned}$$

9780170459242

Question 2

Mathematical induction is used to prove:

$$\frac{2}{1\times 2}+\frac{2}{2\times 3}+\frac{2}{3\times 4}+\cdots+\frac{2}{n\times(n+1)}=\frac{2n}{n+1}$$

for all positive integers n.

Which of the following is the correct expression for the start of the proof for $n = k + 1$?

A $\text{LHS} = \frac{2}{1\times 2}+\frac{2}{2\times 3}+\frac{2}{3\times 4}+\cdots+\frac{2}{k\times(k+1)}$
$= \frac{2k}{k+1}+\frac{2}{k+1}$

B $\text{LHS} = \frac{2}{1\times 2}+\frac{2}{2\times 3}+\frac{2}{3\times 4}+\cdots+\frac{2}{(k+1)}+\frac{2}{(k+2)}$
$= \frac{2k}{k+1}+\frac{2}{k+2}$

C $\text{LHS} = \frac{2}{1\times 2}+\frac{2}{2\times 3}+\frac{2}{3\times 4}+\cdots+\frac{2}{k\times(k+1)}+\frac{2}{(k+1)(k+2)}$
$= \frac{2k}{k+1}+\frac{2}{(k+1)(k+2)}$

D $\text{LHS} = \frac{2}{1\times 2}+\frac{2}{2\times 3}+\frac{2}{3\times 4}+\cdots+\frac{2}{k\times(k+1)}+\frac{2}{(k+1)(k+3)}$
$= \frac{2k}{k+1}+\frac{2}{(k+1)(k+3)}$

Question 3

Finn made an error proving that $3^{2n} - 1$ is divisible by 8 (where n is an integer greater than 0), using mathematical induction. Part of the proof is shown below.

Step 2: Assume the result is true for $n = k$.

$3^{2n} - 1 = 8P$, where P is an integer. Line 1

Hence, $3^{2k} = 8P + 1$.

Step 3: To prove the result is true for $n = k + 1$.

$3^{2(k+1)} - 1 = 8Q$, where Q is an integer. Line 2

$\text{LHS} = 3^{2(k+1)} - 1$
$= 3^{2k} \times 3^2 - 1$
$= (8P + 1) \times 3^2 - 1$ Line 3
$= 72P + 1 - 1$ Line 4
$= 72P$
$= 8 \times 9P$
$= 8Q$
$= \text{RHS}$

In which line did Finn make an error?

A Line 1

B Line 2

C Line 3

D Line 4

Practice set 2

Short-answer questions

Solutions start on page 12.

Question 1 (3 marks) ©NESA 2020 HSC EXAM, QUESTION 12(a)

Use the principle of mathematical induction to show that for all integers $n \geq 1$, 3 marks

$$1 \times 2 + 2 \times 5 + 3 \times 8 + \cdots + n(3n - 1) = n^2(n + 1).$$

Question 2 (2 marks)

Can mathematical induction be used to prove that $\cos(\pi - x) = \cos x$ for all x? 2 marks
Give a reason for your answer.

Question 3 (4 marks)

Determine a formula for the following sum to n terms, and use mathematical induction to prove the formula. 4 marks

$$1 + 4 + 7 + 10 + \ldots$$

Question 4 (3 marks)

Prove by mathematical induction that 3 marks

$$\frac{1}{x} + \frac{1}{x^2} + \frac{1}{x^3} + \cdots + \frac{1}{x^n} = \frac{1}{x - 1} - \frac{1}{x^n(x - 1)}$$

for all positive integers n, $x \neq 0, 1$.

Question 5 (3 marks)

Use mathematical induction to prove that $3^{4n} - 1$ is a multiple of 80 for all positive integers n. 3 marks

Question 6 (3 marks)

Use mathematical induction to prove that $3^{3n} + 2^{n+2}$ is divisible by 5 for all positive integers n. 3 marks

Question 7 (3 marks)

Prove by mathematical induction that 3 marks

$$2 \times 1! + 5 \times 2! + 10 \times 3! + \cdots + (n^2 + 1)n! = n(n + 1)!,$$

where $n! = 1 \times 2 \times 3 \times \cdots \times (n - 1) \times n$.

Question 8 (3 marks) ©NESA 1990 HSC EXAM, QUESTION 7(a)

Use mathematical induction to prove that, for every positive integer n, 3 marks

$$13 \times 6^n + 2$$

is divisible by 5.

Question 9 (2 marks)

Jake worked tirelessly to prove that a statement was indeed true for $n = 1, 2, 3$ and 4.

Has Jake proved that the statement is true for all positive integers n by the method of mathematical induction? Explain your response. 2 marks

Question 10 (3 marks)

Prove that $15^n + 2^{3n} - 2$ is a multiple of 7 if n is odd. 3 marks

PRACTICE SET 2

Question 11 (3 marks)

By the process of mathematical induction, show that: 3 marks

$$\log 2 + \log\left(\frac{3}{2}\right) + \log\left(\frac{4}{3}\right) + \cdots + \log\left(\frac{n+1}{n}\right) = \log(n+1)$$

for all integers $n \geq 1$.

Question 12 (3 marks)

Use mathematical induction to prove that $x^n - 1$ is divisible by $x - 1$ for all positive integers n. 3 marks

Question 13 (3 marks)

The sum of the squares of two consecutive integers is given by $S_n = n^2 + (n+1)^2$.

Prove by mathematical induction whether S_n is divisible by 4 for all integers $n \geq 1$. 3 marks

Question 14 (3 marks)

Prove by mathematical induction: 3 marks

$$a + ar + ar^2 + ar^3 + \cdots + ar^{n-1} = \frac{a(r^n - 1)}{r - 1}$$

for all integers $n \geq 1$.

Question 15 (3 marks)

The sum of the cubes of three consecutive integers is given by

$$S_n = (n-1)^3 + n^3 + (n+1)^3.$$

Prove by mathematical induction that S_n is divisible by 9 for all integers $n \geq 1$. 3 marks

In your proof, you may use the formulas:

$$(a+b)^3 = a^3 + 3ab^2 + 3a^2b + b^3 \text{ and } (a-b)^3 = a^3 - 3ab^2 + 3a^2b - b^3.$$

Question 16 (3 marks)

Prove whether this formula is true for all positive integers n by mathematical induction. 3 marks

$$1^2 + 2^2 + 3^2 + \cdots + n^2 = \left[\frac{1}{2}n(n+1)\right]^2$$

Question 17 (3 marks)

A sequence of numbers is defined by $u_1 = \frac{1}{3}$ and, if n is any positive integer,

$$u_{n+1} = \frac{1 + 3u_n}{3 + u_n}.$$

a Find u_2. 1 mark

b Prove by induction: $u_n = \dfrac{2^n - 1}{2^n + 1}$. 2 marks

Question 18 (3 marks)

One of the functions below is the sum of the series for all positive integers n:

$$1 \times 2 + 3 \times 4 + 5 \times 6 + (2n-1)(2n).$$

State with justification which function is the sum and prove that it is true by mathematical induction. 3 marks

$$f(n) = \frac{1}{3}n(n+1)(4n-1)$$

$$g(n) = \frac{1}{3}(n+1)(4n-1)$$

9780170459242

Practice set 1

Worked solutions

1 B

Step 2: Assume the result is true for $n = k$.

$$1^2 + 3^2 + 5^2 + \cdots + (2k-1)^2 = \frac{1}{3}\boldsymbol{k}(2k+1)(2k-1)$$

2 C

Step 2: Assume the result is true for $n = k$.

$$\frac{2}{1\times 2} + \frac{2}{2\times 3} + \frac{2}{3\times 4} + \cdots + \frac{2}{k\times(k+1)} = \frac{2k}{k+1}$$

To prove the result is true for $n = k + 1$:

$$\frac{2}{1\times 2} + \frac{2}{2\times 3} + \frac{2}{3\times 4} + \cdots + \frac{2}{k\times(k+1)} + \frac{2}{(k+1)(k+2)} = \frac{2(k+1)}{(k+2)}$$

$$\begin{aligned}\text{LHS} &= \frac{2}{1\times 2} + \frac{2}{2\times 3} + \frac{2}{3\times 4} + \cdots + \frac{2}{k\times(k+1)} + \frac{2}{(k+1)(k+2)} \\ &= \frac{2k}{k+1} + \frac{2}{(k+1)(k+2)}\end{aligned}$$

3 D

$72P + 1 - 1$ Line 4 contains the error.

It should say '$= 72P \mathbf{+ 9} - 1$'.

Practice set 2

Worked solutions

Question 1

Step 1: Prove true for $n = 1$ (that is, $T_1 = S_1$).

$$T_1 = 1 \times 2 \\ = 2$$

$$S_1 = 1^2(1 + 1) \\ = 2 \\ = T_1$$

So the statement is true for $n = 1$.

Step 2: Assume the statement is true for $n = k$.

$$1 \times 2 + 2 \times 5 + 3 \times 8 + \cdots + k(3k - 1) = k^2(k + 1)$$

Step 3: Prove the statement is true for $n = k + 1$.

$$\text{RTP } S_{k+1} = (k + 1)^2[(k + 1) + 1] \\ = (k + 1)^2(k + 2)$$

$$\begin{aligned} S_{k+1} &= S_k + T_{k+1} \\ &= k^2(k + 1) + (k + 1)[3(k + 1) - 1] \\ &= k^2(k + 1) + (k + 1)(3k + 2) \\ &= (k + 1)(k^2 + 3k + 2) \\ &= (k + 1)(k + 1)(k + 2) \\ &= (k + 1)^2(k + 2) \end{aligned}$$

Step 4: Conclusion

The statement is true for all integers $n \geq 1$ by mathematical induction.

Question 2

As it states for all values of x, this means it is not just integer values and it could also be positive or negative; so unless x is defined for integer values only, we cannot use mathematical induction.

Question 3

$$\begin{aligned} S_n &= \frac{n}{2}[2a + (n - 1)d] \\ &= \frac{n}{2}[2 + (n - 1)3] \\ &= \frac{n}{2}[2 + 3n - 3] \\ &= \frac{1}{2}n(3n - 1) \end{aligned}$$

$$\begin{aligned} T_n &= a + (n - 1)d \\ &= 1 + (n - 1)3 \\ &= 1 + 3n - 3 \\ &= 3n - 2 \end{aligned}$$

Step 1: Prove true for $n = 1$ (that is, $T_1 = S_1$).

$$T_1 = 3 \times 1 - 2 \\ = 1$$

$$S_1 = \frac{1}{2}(1)(3(1) - 1) \\ = 1 \\ = T_1$$

True for $n = 1$.

Step 2: Assume the statement is true for $n = k$.

$$1 + 4 + 7 + 10 + \cdots + (3k - 2) = \frac{1}{2}k(3k - 1)$$

Step 3: Prove the statement is true for $n = k + 1$.

$$\text{RTP } S_{k+1} = \tfrac{1}{2}(k + 1)(3(k + 1) - 1) \\ = \tfrac{1}{2}(k + 1)(3k + 2)$$

$$\begin{aligned} S_{k+1} &= S_k + T_{k+1} \\ &= \tfrac{1}{2}k(3k - 1) + (3(k + 1) - 2) \\ &= \tfrac{1}{2}k(3k - 1) + (3k + 1) \\ &= \tfrac{1}{2}[k(3k - 1) + 2(3k + 1)] \\ &= \tfrac{1}{2}[3k^2 - k + 6k + 2] \\ &= \tfrac{1}{2}[3k^2 + 5k + 2] \\ &= \tfrac{1}{2}(3k + 2)(k + 1) \\ &= \tfrac{1}{2}(k + 1)(3k + 2) \end{aligned}$$

Step 4: Conclusion

The statement is true for all positive integers n by mathematical induction.

Question 4

Step 1: Prove true for $n = 1$ (that is, $T_1 = S_1$).

$$T_1 = \frac{1}{x}$$

$$S_1 = \frac{1}{x-1} - \frac{1}{x^1(x-1)}$$
$$= \frac{x}{x(x-1)} - \frac{1}{x(x-1)}$$
$$= \frac{x-1}{x(x-1)}$$
$$= \frac{1}{x}$$
$$= T_1$$

So the statement is true for $n = 1$.

Step 2: Assume the statement is true for $n = k$.

$$\frac{1}{x} + \frac{1}{x^2} + \frac{1}{x^3} + \cdots + \frac{1}{x^k} = \frac{1}{x-1} - \frac{1}{x^k(x-1)}$$

Step 3: Prove the statement is true for $n = k + 1$.

$$\text{RTP } S_{k+1} = \frac{1}{x-1} - \frac{1}{x^{k+1}(x-1)}$$

$$S_{k+1} = S_k + T_{k+1}$$
$$= \frac{1}{x-1} - \frac{1}{x^k(x-1)} + \frac{1}{x^{k+1}}$$
$$= \frac{1}{x-1} - \frac{x}{x^{k+1}(x-1)} + \frac{x-1}{x^{k+1}(x-1)}$$

Hint
Common denominators of $x^{k+1}(x - 1)$.

$$= \frac{1}{x-1} - \frac{1}{x^{k+1}(x-1)}$$

Step 4: Conclusion

The statement is true for all positive integers n by mathematical induction.

Question 5

Step 1: Prove true for $n = 1$ (multiple of 80).

$3^{4(1)} - 1 = 80$, which is divisible by 80.
So the statement is true for $n = 1$.

Step 2: Assume the statement is true for $n = k$.

Divide by 80 and then use arrangement in Step 3.

$$\frac{3^{4k} - 1}{80} = M$$
$$3^{4k} - 1 = 80M$$
$$3^{4k} = 80M + 1 \quad [*]$$

Step 3: Prove the statement is true for $n = k + 1$.

Start by substituting $n = k + 1$:

$$3^{4(k+1)} - 1$$
$$= 3^{4k+4} - 1$$
$$= 3^{4k} \times 3^4 - 1$$
$$= (80M + 1)81 - 1 \quad \text{using } [*]$$
$$= 6480M + 81 - 1$$
$$= 6480M + 80$$
$$= 80(81M + 1)$$

Step 4: Conclusion

The statement is true for all positive integers n by mathematical induction.

Question 6

Step 1: Prove true for $n = 1$ (multiple of 5).

$$3^{3(1)} + 2^{1+2} = 35$$
$$= 5 \times 7$$

So the statement is true for $n = 1$.

Step 2: Assume the statement is true for $n = k$.

Divide by 5 and then use arrangement in Step 3.

$$\frac{3^{3k} + 2^{k+2}}{5} = M$$
$$3^{3k} + 2^{k+2} = 5M$$
$$3^{3k} = 5M - 2^{k+2} \quad [*]$$

Step 3: Prove the statement is true for $n = k + 1$.

Start by substituting $n = k + 1$:

$$3^{3(k+1)} + 2^{(k+1)+2}$$
$$= 3^{3k+3} + 2^{k+2+1}$$
$$= 3^{3k} \times 3^3 + 2^{k+2} \times 2^1$$
$$= 27(5M - 2^{k+2}) + 2 \times 2^{k+2} \quad \text{using } [*]$$
$$= 135M - 27 \times 2^{k+2} + 2 \times 2^{k+2}$$
$$= 135M - 25 \times 2^{k+2}$$
$$= 5(27M - 5 \times 2^{k+2})$$

Step 4: Conclusion

The statement is true for all positive integers n by mathematical induction.

WORKED SOLUTIONS

Question 7

Step 1: Prove true for $n = 1$ (that is, $T_1 = S_1$).

$T_1 = 2 \times 1!$
$= 2$

$S_1 = 1(1 + 1)!$
$= 2$
$= T_1$

So the statement is true for $n = 1$.

Step 2: Assume the statement is true for $n = k$.

$2 \times 1! + 5 \times 2! + 10 \times 3! + \cdots + (k^2 + 1)k! = k(k + 1)!$

Step 3: Prove the statement is true for $n = k + 1$.

RTP $S_{k+1} = (k + 1)((k + 1) + 1)!$
$= (k + 1)(k + 2)!$

$$\begin{aligned} S_{k+1} &= S_k + T_{k+1} \\ &= k(k+1)! + ((k+1)^2 + 1)(k+1)! \\ &= k(k+1)! + (k^2 + 2k + 1 + 1)(k+1)! \\ &= k(k+1)! + (k^2 + 2k + 2)(k+1)! \\ &= (k+1)!(k^2 + 2k + 2 + k) \\ &= (k+1)!(k^2 + 3k + 2) \\ &= (k+1)!(k+2)(k+1) \\ &= (k+2)!(k+1) \\ &= (k+1)(k+2)! \end{aligned}$$

Step 4: Conclusion

The statement is true for all positive integers n by mathematical induction.

Question 8

Step 1: Prove true for $n = 1$ (multiple of 5).

$13 \times 6^1 + 2 = 80$
$= 16 \times 5$, which is divisible by 5

So the statement is true for $n = 1$.

Step 2: Assume the statement is true for $n = k$.

Divide by 5 and then use arrangement in Step 3.

$\frac{13 \times 6^k + 2}{5} = M$

$13 \times 6^k + 2 = 5M$

$13 \times 6^k = 5M - 2$ [*]

Step 3: Prove statement is true for $n = k + 1$.

Start by substituting $n = k + 1$:

$13 \times 6^{k+1} + 2$
$= 13 \times 6^k \times 6^1 + 2$
$= 6(5M - 2) + 2$ using [*]
$= 30M - 12 + 2$
$= 30M - 10$
$= 5(16M - 2)$

Step 4: Conclusion

The statement is true for all positive integers n by mathematical induction.

Question 9

Jake has *not* proved the statement is true by mathematical induction because the statement may not be true for $n \geq 5$.

Question 10

Step 1: Prove true for $n = 1$ (multiple of 7).

$15^1 + 2^{3(1)} - 2 = 21$
$= 7 \times 3$, which is divisible by 7.

So the statement is true for $n = 1$.

Step 2: Assume the statement is true for $n = k$.

Divide by 7 and then use arrangement in Step 3.

$\frac{15^k + 2^{3k} - 2}{7} = M$

$15^k + 2^{3k} - 2 = 7M$

$15^k = 7M - 2^{3k} + 2$ [*]

Step 3: Prove true for $n = k + 2$ (n is odd).

Substitute $n = k + 2$:

$15^{k+2} + 2^{3(k+2)} - 2$
$= 15^k \times 15^2 + 2^{3k+6} - 2$
$= 15^k \times 15^2 + 2^{3k} \times 2^6 - 2$
$= 225(7M - 2^{3k} + 2) + 2^{3k} \times 64 - 2$ from [*]
$= 1575M - 225 \times 2^{3k} + 450 + 64 \times 2^{3k} - 2$
$= 1575M - 161 \times 2^{3k} + 448$
$= 7(225M - 23 \times 2^{3k} + 64)$

Step 4: Conclusion

The statement is true for all odd integers $n \geq 1$ by mathematical induction.

Question 11

Step 1: Prove true for $n = 1$ (that is, $T_1 = S_1$).

$T_1 = \log 2$

$$\begin{aligned} S_1 &= \log(1+1) \\ &= \log 2 \\ &= T_1 \end{aligned}$$

So the statement is true for $n = 1$.

Step 2: Assume the statement is true for $n = k$.

$$\log 2 + \log\left(\frac{3}{2}\right) + \log\left(\frac{4}{3}\right) + \cdots + \log\left(\frac{k+1}{k}\right) = \log(k+1)$$

Step 3: Prove the statement is true for $n = k + 1$.

$$\begin{aligned} \text{RTP } S_{k+1} &= \log((k+1)+1) \\ &= \log(k+2) \end{aligned}$$

$$\begin{aligned} S_{k+1} &= S_k + T_{k+1} \\ &= \log(k+1) + \log\left(\frac{(k+1)+1}{(k+1)}\right) \\ &= \log(k+1) + \log\left(\frac{k+2}{k+1}\right) \\ &= \log(k+1) + \log(k+2) - \log(k+1) \\ &= \log(k+2) \end{aligned}$$

Step 4: Conclusion

The statement is true for all integers $n \geq 1$ by mathematical induction.

Question 12

Step 1: Prove true for $n = 1$ (multiple of $x - 1$).

$x^1 - 1 = x - 1$, which is divisible by $x - 1$.

So the statement is true for $n = 1$.

Step 2: Assume the statement is true for $n = k$.

Divide by $x - 1$ and then use arrangement in Step 3.

$$\begin{aligned} \frac{x^k - 1}{x - 1} &= M \\ x^k - 1 &= M(x-1) \\ x^k &= M(x-1) + 1 \quad [*] \end{aligned}$$

Step 3: Prove the statement is true for $n = k + 1$.

Substitute $n = k + 1$:

$$\begin{aligned} x^{k+1} - 1 &= x^1 \times x^k - 1 \\ &= x(M(x-1)+1) - 1 \quad \text{by } [*] \\ &= x(Mx - M + 1) - 1 \\ &= Mx^2 - Mx + x - 1 \\ &= Mx(x-1) + (x-1) \\ &= (x-1)(Mx+1) \end{aligned}$$

Step 4: Conclusion

The statement is true for all positive integers n by mathematical induction.

Question 13

Prove true for $n = 1$ (multiple of 4).

$$\begin{aligned} S_1 &= 1^2 + (1+1)^2 \\ &= 5, \text{ which is not divisible by } 4 \end{aligned}$$

So the statement is not true for $n = 1$.

Conclusion:

The statement is not true because it fails the process of mathematical induction at $n = 1$.

Question 14

Step 1: Prove true for $n = 1$ (that is, $T_1 = S_1$).

$T_1 = a$

$$\begin{aligned} S_1 &= \frac{a(r^1 - 1)}{r - 1} \\ &= \frac{a(r-1)}{r-1} \\ &= a \\ &= T_1 \end{aligned}$$

So the statement is true for $n = 1$.

Step 2: Assume the statement is true for $n = k$.

$$a + ar^1 + ar^2 + ar^3 + \cdots + ar^{k-1} = \frac{a(r^k - 1)}{r - 1}$$

Step 3: Prove the statement is true for $n = k + 1$.

$$\text{RTP } S_{k+1} = \frac{a(r^{k+1} - 1)}{r - 1}:$$

$$\begin{aligned} S_{k+1} &= S_k + T_{k+1} \\ &= \frac{a(r^k - 1)}{r - 1} + ar^{(k+1)-1} \\ &= \frac{a(r^k - 1)}{r - 1} + \frac{ar^k(r-1)}{r-1} \\ &= \frac{ar^k - a + ar^{k+1} - ar^k}{r - 1} \\ &= \frac{ar^{k+1} - a}{r - 1} \\ &= \frac{a(r^{k+1} - 1)}{r - 1} \end{aligned}$$

Step 4: Conclusion

The statement is true for all integers $n \geq 1$ by mathematical induction.

WORKED SOLUTIONS

Question 15

Step 1: Prove true for $n = 1$ (multiple of 9).

$(1-1)^3 + 1^3 + (1+1)^3 = 9$, which is divisible by 9.

So the statement is true for $n = 1$.

Step 2: Assume the statement is true for $n = k$.

Divide by 9 and then use arrangement in Step 3.

$$\frac{(k-1)^3 + k^3 + (k+1)^3}{9} = M$$

$$(k-1)^3 + k^3 + (k+1)^3 = 9M$$

$$k^3 + (k+1)^3 = 9M - (k-1)^3 \quad [*]$$

Step 3: Prove the statement is true for $n = k + 1$.

Substitute $n = k + 1$.

$$((k+1)-1)^3 + (k+1)^3 + ((k+1)+1)^3$$

$$= (k)^3 + (k+1)^3 + (k+2)^3$$

$$= 9M - (k-1)^3 + (k+2)^3 \quad \text{from } [*]$$

$$= 9M - (k^3 - 3k^2 + 3k - 1) + (k^3 + 6k^2 + 12k + 8)$$

$$= 9M - k^3 + 3k^2 - 3k + 1 + k^3 + 6k^2 + 12k + 8$$

$$= 9M + 9k^2 + 9k + 9$$

$$= 9(M + k^2 + k + 1)$$

Step 4: Conclusion

The statement is true for all integers $n \geq 1$ by mathematical induction.

Question 16

Step 1: Prove true for $n = 1$ (that is, $T_1 = S_1$).

$$T_1 = 1^2$$
$$= 1$$

$$S_1 = \left[\tfrac{1}{2}(1)(1+1)\right]^2$$
$$= 1$$
$$= T_1$$

So the statement is true for $n = 1$.

Step 2: Assume the statement is true for $n = k$.

$$1^2 + 2^2 + 3^2 + \cdots + k^2 = \left[\tfrac{1}{2}k(k+1)\right]^2$$

Step 3: Prove the statement is true for $n = k + 1$.

$$\text{RTP } S_{k+1} = \left[\tfrac{1}{2}(k+1)((k+1)+1)\right]^2$$
$$= \left[\tfrac{1}{2}(k+1)(k+2)\right]^2$$

$$S_{k+1} = S_k + T_{k+1}$$

$$= \left[\frac{1}{2}k(k+1)\right]^2 + (k+1)^2$$

$$= \frac{1}{4}k^2(k+1)^2 + (k+1)^2$$

$$= (k+1)^2\left(\frac{1}{4}k^2 + 1\right)$$

$$= \frac{1}{4}(k+1)^2(k^2 + 4)$$

$$\neq \text{RTP}$$

Step 4: Conclusion

The statement is false because it fails the induction process at the induction step.

Question 17

a $u_2 = u_{1+1}$

$= \frac{1+3u_1}{3+u_1}$

$= \frac{1+3\left(\frac{1}{3}\right)}{3+\left(\frac{1}{3}\right)}$

$= \frac{3}{5}$

b Step 1: Prove true for $n = 1$ (that is, $T_1 = u_1$).

$T_1 = \frac{1}{3}$

$u_1 = \frac{2^{(1)}-1}{2^{(1)}+1}$

$= \frac{1}{3}$

$= T_1$

So the statement is true for $n = 1$.

Step 2: Assume the statement is true for $n = k$.

$u_k = \frac{2^k-1}{2^k+1}$

Step 3: Prove the statement is true for $n = k + 1$.

RTP $u_{k+1} = \frac{2^{k+1}-1}{2^{k+1}+1}$

$u_{k+1} = \frac{1+3u_k}{3+u_k}$

$= \frac{1+3\left(\frac{2^k-1}{2^k+1}\right)}{3+\frac{2^k-1}{2^k+1}}$

$= \frac{1(2^k+1)+3\times 2^k-3}{3(2^k+1)+2^k-1}$

$= \frac{2^k+1+3\times 2^k-3}{3\times 2^k+3+2^k-1}$

$= \frac{4\times 2^k-2}{4\times 2^k+2}$

$= \frac{2(2\times 2^k-1)}{2(2\times 2^k+1)}$

$= \frac{2\times 2^k-1}{2\times 2^k+1}$

$= \frac{2^{k+1}-1}{2^{k+1}+1}$ as required

Step 4: Conclusion

The statement is true for all positive integers n by mathematical induction.

Question 18

$S_1 = 1 \times 2$

$= 2$

$S_2 = 1 \times 2 + 3 \times 4$

$= 14$

$f(1) = \frac{1}{3}(1)(1+1)(4(1)-1)$

$= 2$

$f(2) = \frac{1}{3}(2)(2+1)(4(2)-1)$

$= 14$

$g(1) = \frac{1}{3}(1+1)(4(1)-1)$

$= 2$

$g(2) = \frac{1}{3}(2+1)(4(2)-1)$

$= 7$

$\neq 14$

So $f(n) = \frac{1}{3}n(n+1)(4n-1)$ could be the sum.

Step 1: Prove true for $n = 1$ (that is, $T_1 = S_1$).

Already proved by $f(1)$ above.

So the statement is true for $n = 1$.

Step 2: Assume the statement is true for $n = k$.

$1 \times 2 + 3 \times 4 + 5 \times 6 + (2k-1)(2k)$

$= \frac{1}{3}k(k+1)(4k-1)$

Step 3: Prove true for $n = k + 1$.

RTP $S_{k+1} = \frac{1}{3}(k+1)((k+1)+1)(4(k+1)-1)$

$= \frac{1}{3}(k+1)(k+2)(4k+3)$

$S_{k+1} = S_k + T_{k+1}$

$= \frac{1}{3}k(k+1)(4k-1) + (2(k+1)-1)(2(k+1))$

$= \frac{1}{3}k(k+1)(4k-1) + (2k+1)(2(k+1))$

$= (k+1)[(\frac{1}{3}k(4k-1)) + 2(2k+1)]$

$= (k+1)\frac{1}{3}[(4k^2-k) + 6(2k+1)]$

$= \frac{1}{3}(k+1)[4k^2-k+12k+6]$

$= \frac{1}{3}(k+1)[4k^2+11k+6]$

$= \frac{1}{3}(k+1)(k+2)(4k+3)$ as required

Step 4: Conclusion

The statement is true for all positive integers n by mathematical induction.

HSC exam topic grid (2011–2020)

This table shows the coverage of this topic in past HSC exams by question number. The past exams can be downloaded from the NESA website (www.educationstandards.nsw.edu.au) by selecting 'Year 11 – Year 12', 'HSC exam papers'. NESA marking feedback and guidelines can also be found there.

The new Mathematics Extension 1 course was first examined in 2020. For exams before 2020, select 'Year 11 – Year 12', 'Resources archive', 'HSC exam papers archive'.

	Series proofs	**Divisibility proofs**
2011	6(a)	
2012		12(a)
2013	14(a)	
2014		13(a)
2015	13(c)	
2016	14(a)	
2017		**14(a)**
2018	13(a)	
2019	14(a)	
2020 new course	**12(a)**	

Questions in **bold** can be found in this chapter.

CHAPTER 2
VECTORS

ME-V1 Introduction to vectors 22
V1.1 Introduction to vectors 22
V1.2 Further operations with vectors 24
V1.3 Projectile motion 27

2

VECTORS

Operations with vectors

- Magnitude
- Length and angle
- Adding and subtracting
- Scalar multiplication
- Scalar (dot) product
- Projections

Types of vectors

- Position vector
- Zero vector
- Parallel and perpendicular vectors
- Unit vector

Geometric descriptions

- Column and component form
- The angle between 2 vectors
- Displacement vector
- Geometrical proofs

Projectile motion

- Deriving equations
- Vector functions
- Time equations

Glossary

A+ DIGITAL FLASHCARDS
Revise this topic's key terms and concepts by scanning the QR code or typing the URL into your browser.

https://get.ga/a-hsc-maths-ext-1

acceleration ($\underset{\sim}{a}$)
Change in velocity. On a projectile, it is usually caused by gravity ($\underset{\sim}{a} = -\underset{\sim}{g}$).

Cartesian equation
An equation in terms of x and y on the number plane.

column vector notation
Representation of a vector in the form $\begin{pmatrix} a \\ b \end{pmatrix}$.

component form
Representation of a vector in form $a\underset{\sim}{i} + b\underset{\sim}{j}$, where $\underset{\sim}{i}$ is the unit vector in the x-direction and $\underset{\sim}{j}$ is the unit vector in the y-direction.

displacement ($\underset{\sim}{r}$)
The position of an object at a particular time.

displacement vector
A vector that represents displacement between 2 points on a number plane.

equal vectors
Vectors that have the same magnitude (size) and direction.

gravity
A constant force acting on an object, moving vertically downwards close to Earth's surface (assume that $g = 10\,\text{m}\,\text{s}^{-2}$).

initial position
The starting position of a projectile.

initial velocity
The starting velocity of a projectile, or the speed and direction at which it is launched.

magnitude of a vector
The size or length of a vector, measured as a number. The vector $a\underset{\sim}{i} + b\underset{\sim}{j}$ has magnitude $\sqrt{a^2 + b^2}$.

maximum height
The highest vertical point the projectile can reach above the ground. This height is reached when the projectile's vertical velocity is 0 ($\dot{y} = 0$).

orthogonal
Perpendicular.

parallel vectors
Vectors that have the same or opposite direction.

particle
A point representing an object undergoing motion.

path
The curve traced out by an object as it completes its motion. For a projectile, it is a parabola.

position vector
A vector that starts at the origin, O.

projectile
An object that is launched or dropped under gravity.

range
The horizontal distance between the projection point and where the projectile lands.

scalar
A quantity that has magnitude but no direction. A number, not a vector.

scalar product (or dot product)
The numerical product of two vectors such as $\underset{\sim}{u}$ and $\underset{\sim}{v}$.

$$\underset{\sim}{u} \cdot \underset{\sim}{v} = x_1 x_2 + y_1 y_2$$

$$\underset{\sim}{u} \cdot \underset{\sim}{v} = |\underset{\sim}{u}||\underset{\sim}{v}|\cos\theta$$

scalar projection
The length of the vector projection.

time of flight
The time from the launch of the projectile to when the projectile lands.

unit vector
A vector of magnitude 1. Examples include the vector $\underset{\sim}{i}$ (x-direction) or $\underset{\sim}{j}$ (y-direction).

vector
A quantity with both magnitude and size, with notation $\underset{\sim}{a}$, $\overrightarrow{AB}$ or **a** (bolded).

vector projection
The vector created by projecting one vector onto another vector.

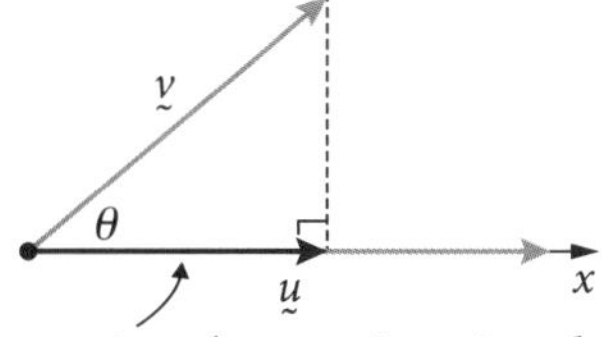

$\text{proj}_{\underset{\sim}{u}}\underset{\sim}{v}$ (vector $\underset{\sim}{v}$ is projected onto vector $\underset{\sim}{u}$)

velocity ($\underset{\sim}{v}$)
The speed of a projectile and the direction it is travelling at a particular time.

Topic summary

Introduction to vectors (ME-V1)

V1.1 Introduction to vectors

Scalars and vectors

A **scalar** is a quantity with magnitude (size) but not direction.

A **vector** is a quantity with both magnitude and direction.

$\underset{\sim}{a}$ or $\overrightarrow{AB}$ or $\mathbf{a}$ is a vector.

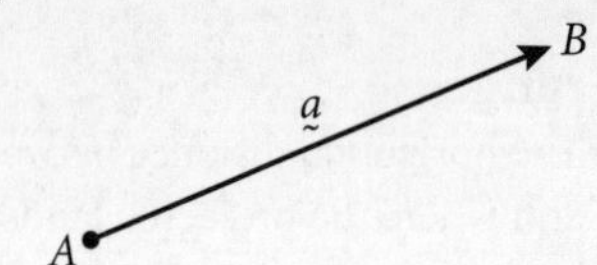

A = tail

B = head

Length or magnitude of $\underset{\sim}{a}$ is $|\underset{\sim}{a}|$.

A **displacement vector** represents **displacement** (movement) between two points on a number plane.

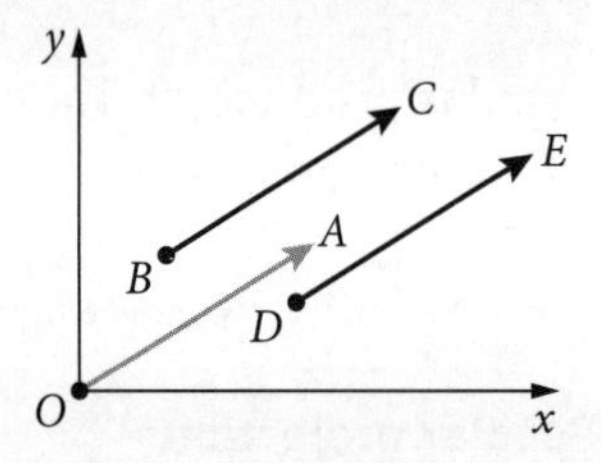

$\overrightarrow{BC}$ and $\overrightarrow{DE}$ are both displacement vectors and can be located anywhere on the Cartesian plane.

$\overrightarrow{OA}$ is a special displacement vector called the **position vector** of A as it starts at the origin.

Parallel vectors have the same or opposite direction.

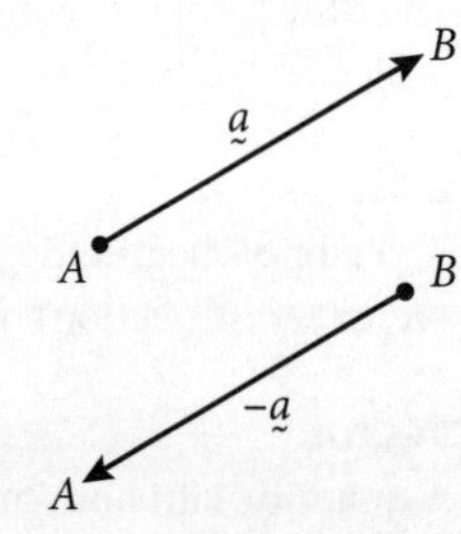

$\overrightarrow{AB}$ and $\overrightarrow{BA}$ are opposite vectors because they have the same magnitude $|\underset{\sim}{a}|$, but opposite direction.

$\overrightarrow{AB} = \underset{\sim}{a}$

$\overrightarrow{BA} = -\underset{\sim}{a}$

$\overrightarrow{OA}$, $\overrightarrow{BC}$ and $\overrightarrow{DE}$ are parallel and have the same magnitude so are **equal vectors**.

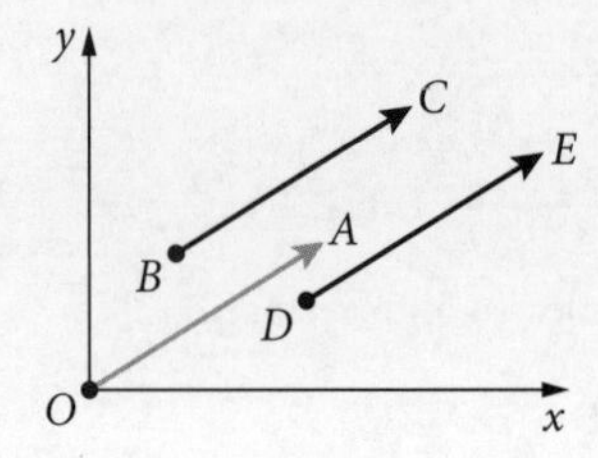

$\underset{\sim}{a}$ and $\underset{\sim}{b}$ are parallel but have different magnitudes.

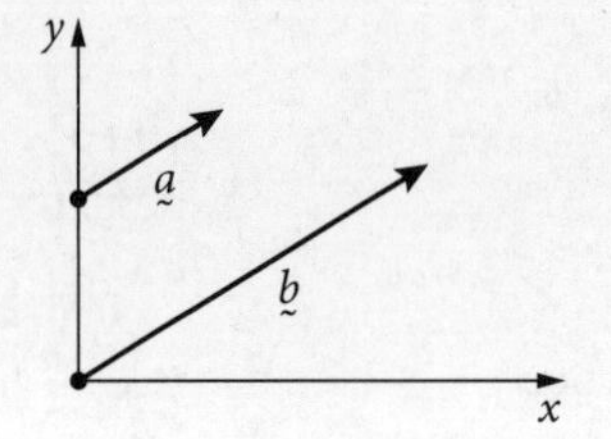

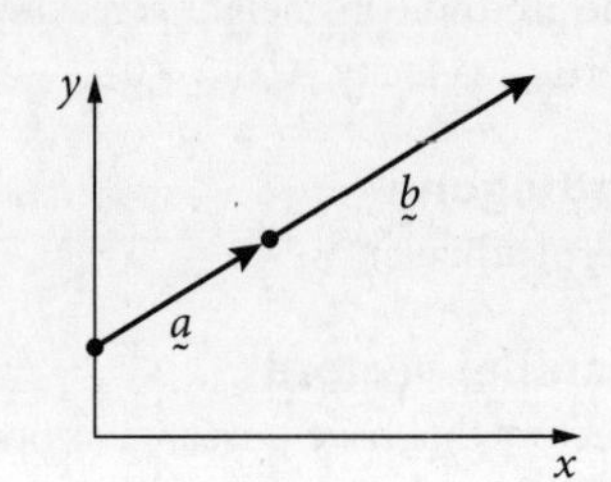

Scalar multiplication

- When multiplying a vector by a scalar, its length is multiplied by this number.

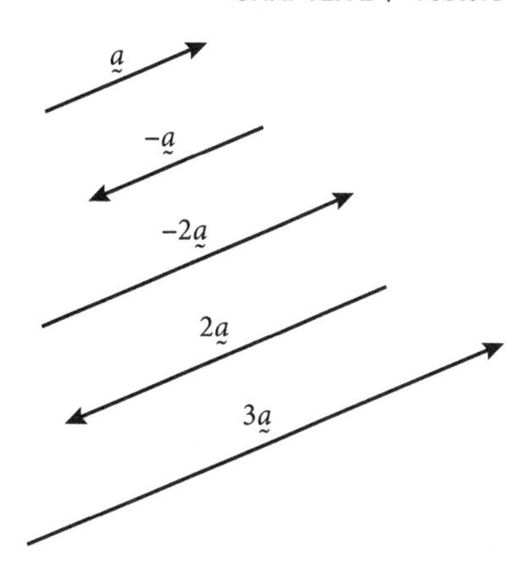

Zero vector

- A zero vector has no length and no direction.
- It is parallel to itself and every other vector.
- It is represented by a point rather than an arrow.

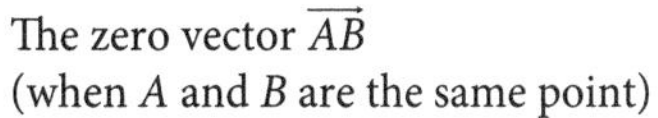

The zero vector $\overrightarrow{AB}$
(when A and B are the same point)

Adding and subtracting vectors

Two vectors can be added together by placing them head to tail. They give a single or *resultant* vector (which is the third side of a triangle).

Triangle law of addition and subtraction

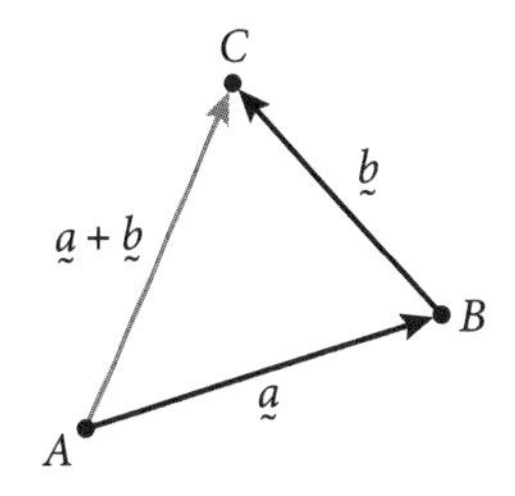

$\overrightarrow{AC} = \overrightarrow{AB} + \overrightarrow{BC}$

Subtracting: $\underset{\sim}{a} - \underset{\sim}{b} = \underset{\sim}{a} + (-\underset{\sim}{b})$

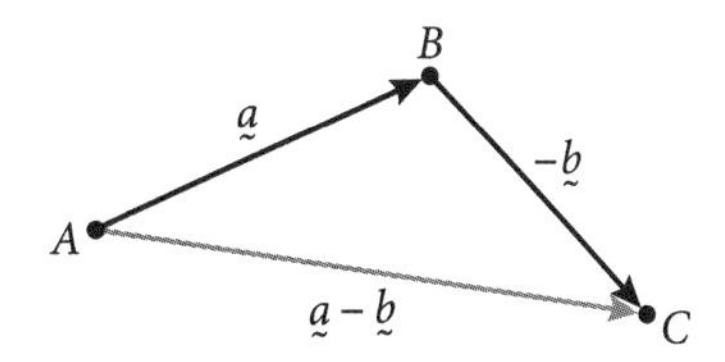

Parallelogram law of addition and subtraction

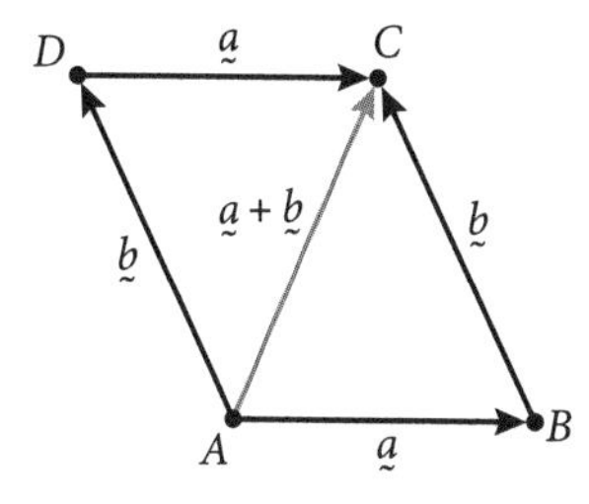

Commutative law: $\underset{\sim}{a} + \underset{\sim}{b} = \underset{\sim}{b} + \underset{\sim}{a}$

Subtracting:

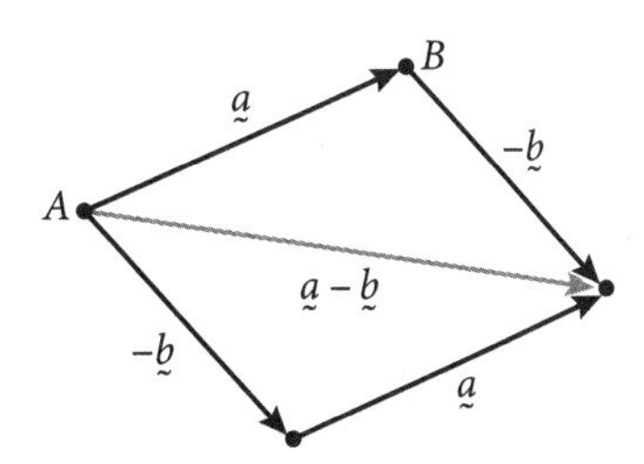

Note: Adding a vector to its negative gives the zero vector.

$$\underset{\sim}{a} + (-\underset{\sim}{a}) = \underset{\sim}{0}$$

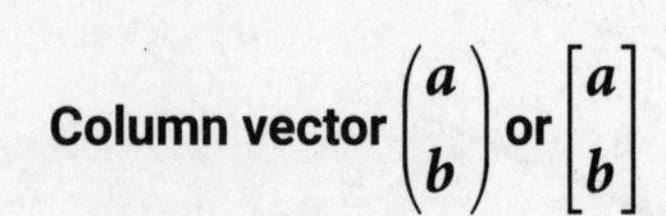

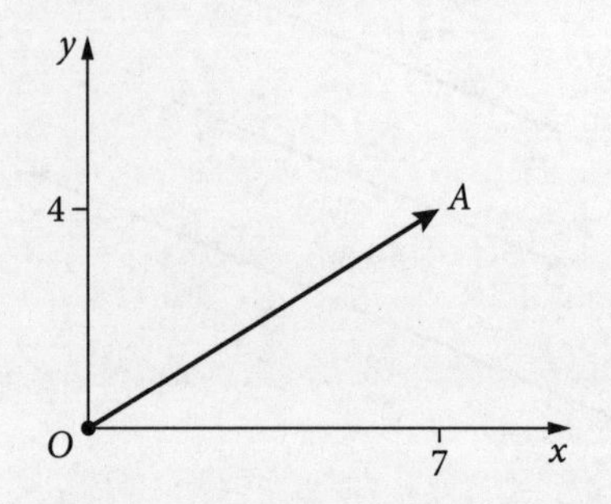

- A is located at the point with coordinates $(7,4)$.
- The vector $\overrightarrow{OA}$ is represented by $\begin{pmatrix} 7 \\ 4 \end{pmatrix}$
- A **position vector** $\begin{pmatrix} a \\ b \end{pmatrix}$ represents a vector from the origin to the point (a, b) on a number plane.
  - If $a > 0$, the direction is to the right.
  - If $a < 0$, the direction is to the left.
  - If $b > 0$, the direction is up.
  - If $b < 0$, the direction is down.

V1.2 Further operations with vectors

Magnitude of a vector

- The **magnitude of a vector** is its length.
- Absolute value symbols are used to represent the magnitude of a vector:

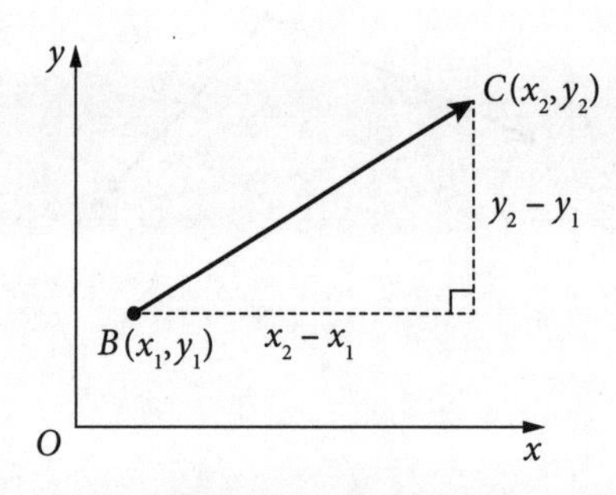

Displacement vector

$$|\overrightarrow{BC}| = \sqrt{(x_2 - x_1)^2 + (y_2 - y_1)^2}$$

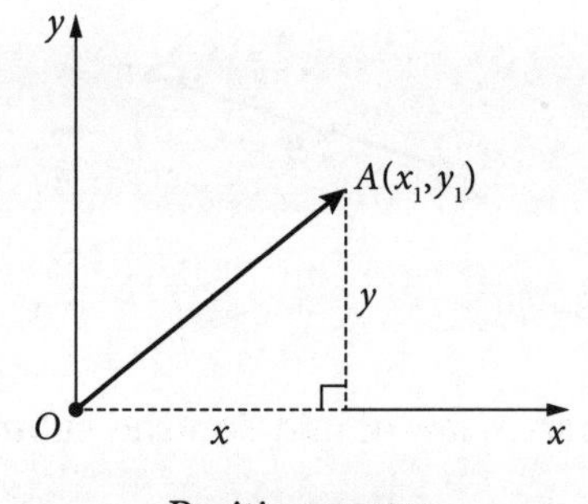

Position vector

$$|\overrightarrow{OA}| = \sqrt{x^2 + y^2}$$

Unit vector, $\hat{\underset{\sim}{a}}$

- The length of a unit vector is $|\hat{\underset{\sim}{a}}| = 1$ unit.
- There are 2 unit vectors parallel to $\underset{\sim}{a}$:
  - Same direction: $\hat{\underset{\sim}{a}} = \dfrac{\underset{\sim}{a}}{|\underset{\sim}{a}|}$
  - Opposite direction: $\hat{\underset{\sim}{a}} = -\dfrac{\underset{\sim}{a}}{|\underset{\sim}{a}|}$

Length and angle

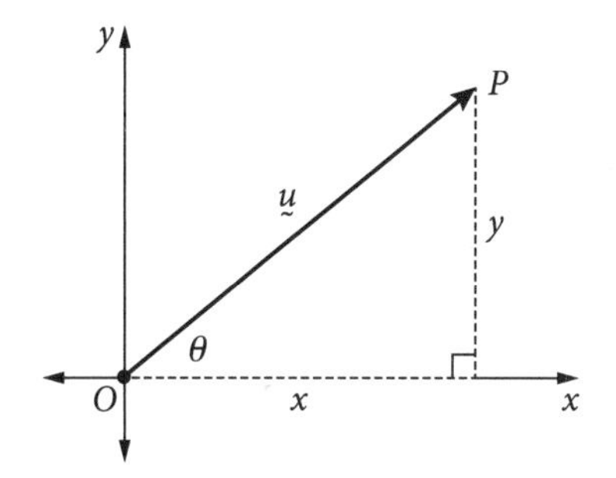

Let $\underset{\sim}{u} = x\underset{\sim}{i} + y\underset{\sim}{j} = \begin{bmatrix} x \\ y \end{bmatrix}$

$\left|\underset{\sim}{u}^2\right| = \sqrt{x^2 + y^2}$ and $\tan\theta = \dfrac{y}{x}$,

where θ is the angle made with the positive x-axis, $0° \le \theta < 360°$.

$x = |\underset{\sim}{u}|\cos\theta$ and $y = |\underset{\sim}{u}|\sin\theta$

Addition and subtraction

- $\underset{\sim}{u} = x_1\underset{\sim}{i} + y_1\underset{\sim}{j}$ and $\underset{\sim}{v} = x_2\underset{\sim}{i} + y_2\underset{\sim}{j}$
- $\underset{\sim}{u} = \begin{bmatrix} x_1 \\ y_1 \end{bmatrix}$ and $\underset{\sim}{v} = \begin{bmatrix} x_2 \\ y_2 \end{bmatrix}$
- $\underset{\sim}{u} + \underset{\sim}{v} = \begin{bmatrix} x_1 \\ y_1 \end{bmatrix} + \begin{bmatrix} x_2 \\ y_2 \end{bmatrix} = \begin{bmatrix} x_1 + x_2 \\ y_1 + y_2 \end{bmatrix}$
- $\underset{\sim}{u} - \underset{\sim}{v} = \begin{bmatrix} x_1 \\ y_1 \end{bmatrix} - \begin{bmatrix} x_2 \\ y_2 \end{bmatrix} = \begin{bmatrix} x_1 - x_2 \\ y_1 - y_2 \end{bmatrix}$

Scalar multiplication

$$\lambda\begin{bmatrix} x_1 \\ y_1 \end{bmatrix} = \begin{bmatrix} \lambda x_1 \\ \lambda y_1 \end{bmatrix}$$

Scalar product (or dot product)

If $\underset{\sim}{a} = x_1\underset{\sim}{i} + y_1\underset{\sim}{j}$ and $\underset{\sim}{b} = x_2\underset{\sim}{i} + y_2\underset{\sim}{j}$, then $\underset{\sim}{a} \cdot \underset{\sim}{b} = x_1 x_2 + y_1 y_2$.

- The scalar (or dot) product multiplies 2 vectors. It applies the directional growth of 1 vector to another. The dot is a raised dot: $\underset{\sim}{a} \cdot \underset{\sim}{b}$.
- The dot product is closely associated with the cosine rule.

  Given $\underset{\sim}{a} = \begin{bmatrix} x_1 \\ y_1 \end{bmatrix}$ and $\underset{\sim}{b} = \begin{bmatrix} x_2 \\ y_2 \end{bmatrix}$ then $\underset{\sim}{a} \cdot \underset{\sim}{b} = x_1x_2 + y_1y_2$

  $$\underset{\sim}{a} \cdot \underset{\sim}{b} = |\underset{\sim}{a}||\underset{\sim}{b}|\cos\theta$$

- If $\underset{\sim}{a}$ or $\underset{\sim}{b}$ is the zero vector, then $\underset{\sim}{a} \cdot \underset{\sim}{b} = 0$.

When working with vectors geometrically, they must share a common tail. If not, move one vector to the tail of the other while maintaining direction and length (magnitude).

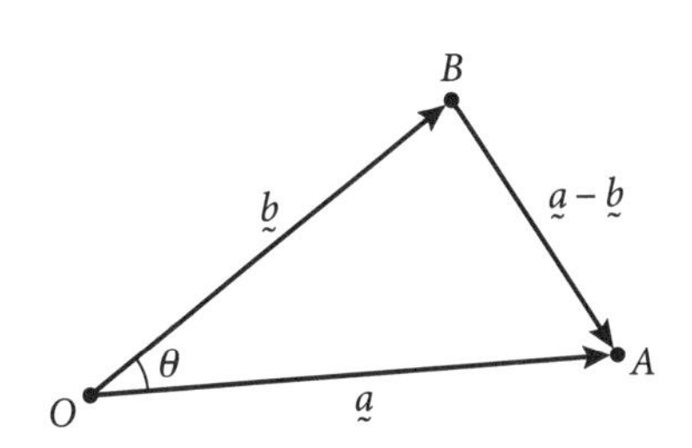

Properties of the scalar product

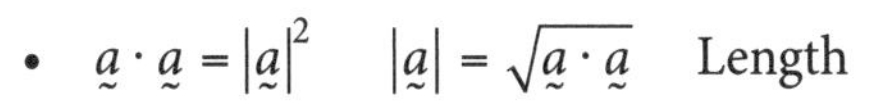

- $\underset{\sim}{a} \cdot \underset{\sim}{a} = |\underset{\sim}{a}|^2 \quad |\underset{\sim}{a}| = \sqrt{\underset{\sim}{a} \cdot \underset{\sim}{a}}$ Length
- $\underset{\sim}{a} \cdot \underset{\sim}{b} = \underset{\sim}{b} \cdot \underset{\sim}{a}$ Commutative law
- $\lambda(\underset{\sim}{a} \cdot \underset{\sim}{b}) = \lambda(\underset{\sim}{a}) \cdot \underset{\sim}{b}$ Associative law
- $\underset{\sim}{a} \cdot (\underset{\sim}{b} + \underset{\sim}{c}) = \underset{\sim}{a} \cdot \underset{\sim}{b} + \underset{\sim}{a} \cdot \underset{\sim}{c}$ Distributive law

Angle between 2 vectors

- The angle between $\underset{\sim}{a}$ and $\underset{\sim}{b}$ = θ.
- $\underset{\sim}{a}$ and $\underset{\sim}{b}$ have the same direction if and only if $\theta = 0°$.
- $\underset{\sim}{a}$ and $\underset{\sim}{b}$ have the opposite direction if and only if $\theta = 180°$.

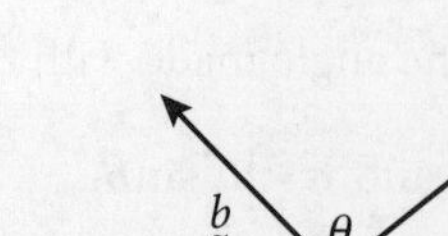

- The angle between $-\underset{\sim}{a}$ and $-\underset{\sim}{b}$ = θ.
- The angle between $\underset{\sim}{a}$ and $-\underset{\sim}{b}$ = $180° - \theta$.
- The angle between $-\underset{\sim}{a}$ and $\underset{\sim}{b}$ = $180° - \theta$.

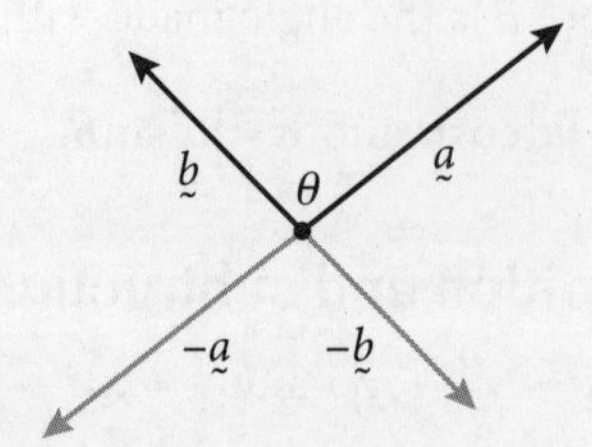

Perpendicular vectors

Two vectors $\underset{\sim}{a}$ and $\underset{\sim}{b}$ are perpendicular (angle between is 90°) if $\underset{\sim}{a} \cdot \underset{\sim}{b} = 0$ ($\cos 90° = 0$).

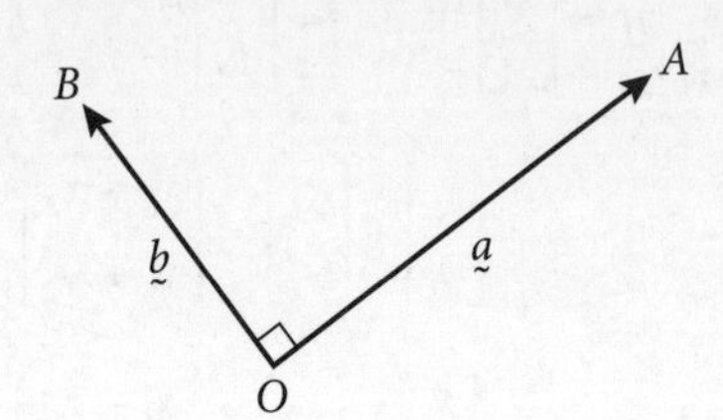

Standard unit vectors

- $\underset{\sim}{i}$ is the unit vector in the positive direction on the x-axis.
  $\underset{\sim}{j}$ is the unit vector in the positive direction on the y-axis.

  $\underset{\sim}{i} = \begin{pmatrix} 1 \\ 0 \end{pmatrix}$ and $\underset{\sim}{j} = \begin{pmatrix} 0 \\ 1 \end{pmatrix}$

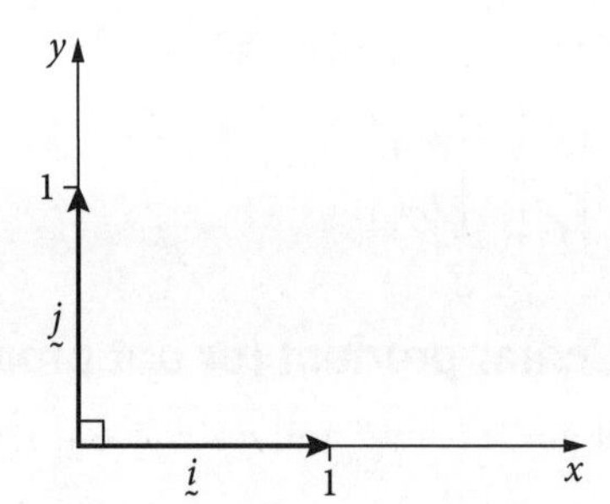

- Any vector can be written as a sum of its components:

  $$\begin{bmatrix} a \\ b \end{bmatrix} = a\underset{\sim}{i} + b\underset{\sim}{j}$$

Geometric proofs: Examples (see questions from Practice set 2 on p.35)

- The diagonals of a parallelogram meet at right angles if and only if it is a rhombus. (Refer to Question 2.)
- The midpoints of the sides of a quadrilateral join to form a parallelogram. (Refer to Question 6.)
- The sum of the squares of the lengths of the diagonals of a parallelogram is equal to the sum of the squares of the lengths of the sides. (Refer to Question 4.)

Projection of a vector

A vector can be projected onto another vector. It is as if a shadow is being cast from a light above. The diagrams show the projection of $\underset{\sim}{v}$ onto $\underset{\sim}{u}$, written $\text{proj}_{\underset{\sim}{u}}\underset{\sim}{v}$, which is another vector, in the same direction as $\underset{\sim}{u}$.

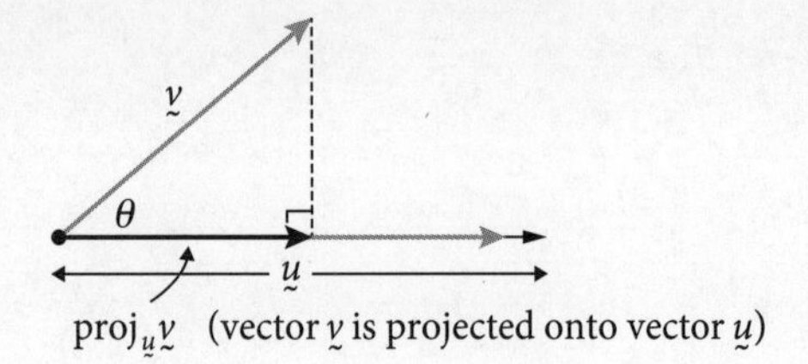

$\text{proj}_{\underset{\sim}{u}}\underset{\sim}{v}$ (vector $\underset{\sim}{v}$ is projected onto vector $\underset{\sim}{u}$)

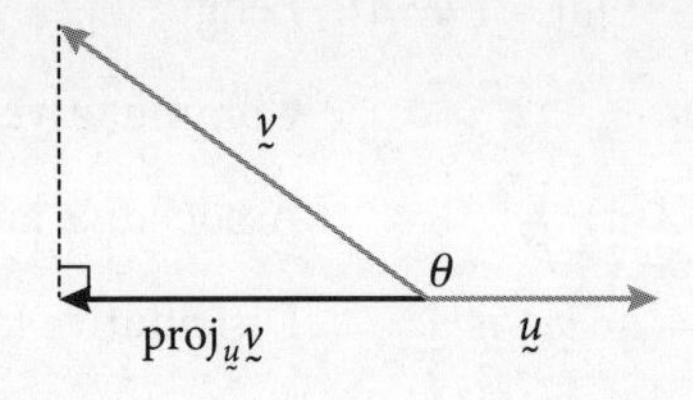

The **scalar projection** of $\underset{\sim}{v}$ onto $\underset{\sim}{u}$ is the *length* of the vector projection:

$$\left|\text{proj}_{\underset{\sim}{u}}\underset{\sim}{v}\right| = \frac{\underset{\sim}{u} \cdot \underset{\sim}{v}}{\left|\underset{\sim}{u}\right|}$$

The **vector projection** of $\underset{\sim}{v}$ onto $\underset{\sim}{u}$ is found by multiplying the scalar projection by the unit vector $\hat{\underset{\sim}{u}} = \frac{\underset{\sim}{u}}{\left|\underset{\sim}{u}\right|}$:

$$\begin{aligned}\text{proj}_{\underset{\sim}{u}}\underset{\sim}{v} &= \left|\text{proj}_{\underset{\sim}{u}}\underset{\sim}{v}\right|\frac{\underset{\sim}{u}}{\left|\underset{\sim}{u}\right|} \\ &= \frac{\underset{\sim}{u} \cdot \underset{\sim}{v}}{\left|\underset{\sim}{u}\right|}\frac{\underset{\sim}{u}}{\left|\underset{\sim}{u}\right|} \\ &= \frac{\underset{\sim}{u} \cdot \underset{\sim}{v}}{\left|\underset{\sim}{u}\right|^2}\underset{\sim}{u}\end{aligned}$$

$$\text{OR} \quad \text{proj}_{\underset{\sim}{u}}\underset{\sim}{v} = \frac{\underset{\sim}{u} \cdot \underset{\sim}{v}}{\underset{\sim}{u} \cdot \underset{\sim}{u}}\underset{\sim}{u}$$

- **Orthogonal projection** is the perpendicular component of the projection of $\underset{\sim}{v}$ onto $\underset{\sim}{u}$; let this be vector $\underset{\sim}{w}$.
- $\underset{\sim}{w}$ is also significant as it is the shortest vector from any point on $\underset{\sim}{u}$ to the tip of vector $\underset{\sim}{v}$.
- $\underset{\sim}{w} = \underset{\sim}{v} - \text{proj}_{\underset{\sim}{u}}\underset{\sim}{v}$ (the vector between $\text{proj}_{\underset{\sim}{u}}\underset{\sim}{v}$ and $\underset{\sim}{v}$)

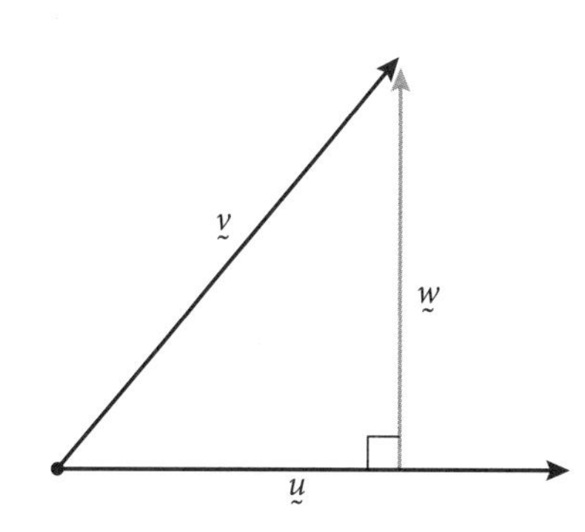

Hint
Orthogonal projection is not in the syllabus.

V1.3 Projectile motion

There are 6 equations of motion (time equations):

- Horizontal: displacement, **velocity**, **acceleration**
- Vertical: displacement, velocity, acceleration

The acceleration equations

- Horizontal component: $\ddot{x} = 0$
- Vertical component: $\ddot{y} = -g$

Unless otherwise stated, we begin with these 2 equations.

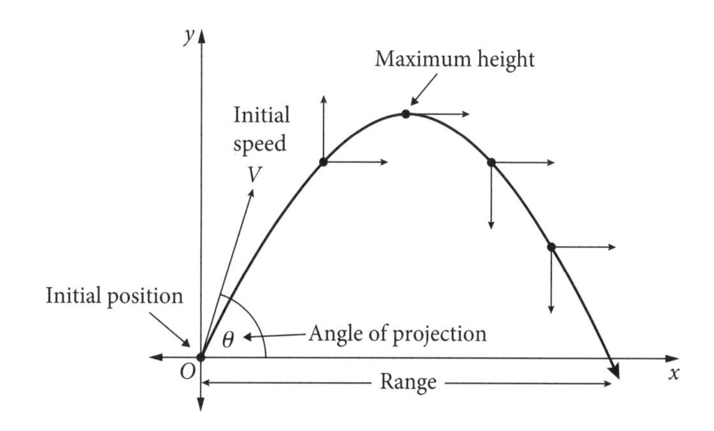

TOPIC SUMMARY

The 6 equations of projectile motion

The **general case** (you need to derive each time unless question gives it):

	Horizontally	Vertically
Acceleration		
	$\ddot{x} = 0$	$\ddot{y} = -g$
Velocity		
	$\dot{x} = \int 0\ dt$ $\dot{x} = c_1$	$\dot{y} = \int -g\ dt$ $\dot{y} = -gt + c_2$
Initial conditions $t = 0$: $\dot{x} = V\cos\theta$, $\dot{y} = V\sin\theta$		
	$V\cos\theta = c_1$ $\therefore \dot{x} = V\cos\theta$	$V\sin\theta = -g(0) + c_2$ $\therefore \dot{y} = -gt + V\sin\theta$
Displacement		
	$x = \int V\cos\theta\ dt$ $x = Vt\cos\theta + c_3$	$y = \int -gt + V\sin\theta\ dt$ $y = \dfrac{-gt^2}{2} + Vt\sin\theta + c_4$
Initial conditions $t = 0$: $x = 0, y = 0$		
	$0 = V(0)\cos\theta + c_3$ $0 = c_3$ So $x = Vt\cos\theta$	$0 = \dfrac{-g(0)^2}{2} + V(0)\sin\theta + c_4$ $0 = c_4$ $\therefore y = \dfrac{-gt^2}{2} + Vt\sin\theta$

Cartesian equation of the path

Initial displacement in horizontal direction:

$$x = Vt\cos\theta$$

$$\frac{x}{V\cos\theta} = t \qquad [1]$$

Initial displacement in vertical direction:

$$y = \frac{-gt^2}{2} + Vt\sin\theta \qquad [2]$$

Substitute [1] into [2]:

$$y = \frac{-g}{2}\left(\frac{x}{V\cos\theta}\right)^2 + V\left(\frac{x}{V\cos\theta}\right)\sin\theta$$

$$y = \frac{-gx^2}{2V^2\cos^2\theta} + \frac{x\sin\theta}{\cos\theta} \qquad \left(\tan\theta = \frac{\sin\theta}{\cos\theta}\right)$$

$$y = \frac{-gx^2}{2V^2}\sec^2\theta + x\tan\theta \qquad \left(\sec^2\theta = \frac{1}{\cos^2\theta}\right)$$

$$\text{OR} \quad y = \frac{-gx^2}{2V^2}(\tan^2\theta + 1) + x\tan\theta \qquad (\text{since } \sec^2\theta = \tan^2\theta + 1)$$

> **Hint**
> When approaching an exam question, consider whether it will be easier for you to keep all constants as pronumerals or if substituting in values will simplify the problem. If the values of pronumerals are simple (for example, integers or basic fractions), it may be better to substitute them in first.

Vector functions for projectile motion

Standard notation for the position vector of a curve is $\underset{\sim}{r}$, then the 3 equations of motion are:

$$\underset{\sim}{a} = \begin{pmatrix} \ddot{x} \\ \ddot{y} \end{pmatrix} = \begin{pmatrix} 0 \\ -g \end{pmatrix}$$

$$\underset{\sim}{v} = \begin{pmatrix} \dot{x} \\ \dot{y} \end{pmatrix} = \begin{pmatrix} V\cos\theta \\ -gt + V\sin\theta \end{pmatrix}$$

$$\underset{\sim}{r} = \begin{pmatrix} x \\ y \end{pmatrix} = \begin{pmatrix} Vt\cos\theta \\ -\frac{g}{2}t^2 + Vt\sin\theta \end{pmatrix}$$

The position of a **particle** can be derived purely from vectors, sometimes producing a simpler proof:

Acceleration: $\underset{\sim}{a} = \begin{pmatrix} 0 \\ -g \end{pmatrix}$

Velocity: $\underset{\sim}{v} = \int \begin{pmatrix} 0 \\ -g \end{pmatrix} dt$

$$= \begin{pmatrix} 0 \\ -g \end{pmatrix} t + c_1$$

When $t = 0$, $\underset{\sim}{v} = \begin{pmatrix} V\cos\theta \\ V\sin\theta \end{pmatrix}$

$$c_1 = \begin{pmatrix} V\cos\theta \\ V\sin\theta \end{pmatrix}$$

$$\therefore \underset{\sim}{v} = \begin{pmatrix} 0 \\ -g \end{pmatrix} t + \begin{pmatrix} V\cos\theta \\ V\sin\theta \end{pmatrix}$$

$$= \begin{pmatrix} V\cos\theta \\ -gt + V\sin\theta \end{pmatrix}.$$

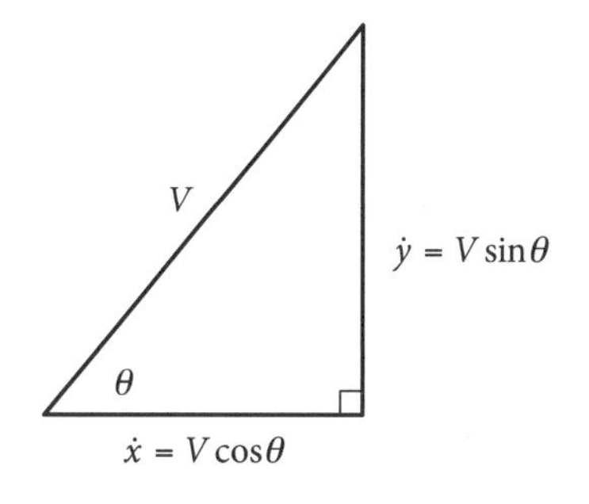

Displacement: $\underset{\sim}{r} = \int \begin{pmatrix} 0 \\ -g \end{pmatrix} t + \begin{pmatrix} V\cos\theta \\ V\sin\theta \end{pmatrix} dt$

$$= \begin{pmatrix} 0 \\ -g \end{pmatrix} \times \frac{t^2}{2} + \begin{pmatrix} V\cos\theta \\ V\sin\theta \end{pmatrix} t + c_2$$

When $t = 0$, $\underset{\sim}{r} = \begin{pmatrix} 0 \\ 0 \end{pmatrix}$

$$c_2 = \begin{pmatrix} 0 \\ 0 \end{pmatrix}$$

$$\therefore \underset{\sim}{r} = \begin{pmatrix} 0 \\ -g \end{pmatrix} \times \frac{t^2}{2} + \begin{pmatrix} V\cos\theta \\ V\sin\theta \end{pmatrix} t$$

$$= \begin{pmatrix} Vt\cos\theta \\ -\frac{gt^2}{2} + Vt\sin\theta \end{pmatrix}.$$

Practice sets tracking grid

Maths is all about repetition, meaning do, do and do again! Each question in the following practice sets, especially the struggle questions (different for everybody!), should be completed at least 3 times correctly. Below is a tracking grid to record your question attempts: ✓ if you answered correctly, ✗ if you didn't.

PRACTICE SET 1: Multiple-choice questions

Question	1st attempt	2nd attempt	3rd attempt	4th attempt	5th attempt
1					
2					
3					
4					
5					
6					
7					
8					
9					
10					
11					
12					
13					
14					
15					
16					
17					
18					
19					
20					

PRACTICE SET 2: Short-answer questions

Question	1st attempt	2nd attempt	3rd attempt	4th attempt	5th attempt
1					
2					
3					
4					
5					
6					
7					
8					
9					
10					
11					
12					
13					
14					
15					
16					
17					
18					
19					
20					

9780170459242

Practice set 1

Multiple-choice questions

Solutions start on page 40.

Question 1

Which of the following is a scalar measurement?

A walking north 6 km

B swimming 2 km

C $\overrightarrow{OA}$

D $\underset{\sim}{v} = 200\underset{\sim}{i}$

Question 2

Which vector is not shown on the diagram?

A $\underset{\sim}{p} = 4\underset{\sim}{i} - 3\underset{\sim}{j}$

B $\underset{\sim}{m} = -6\underset{\sim}{i} + \underset{\sim}{j}$

C $\underset{\sim}{n} = \begin{pmatrix} -1 \\ -4 \end{pmatrix}$

D $\underset{\sim}{q} = \begin{pmatrix} 4 \\ 4 \end{pmatrix}$

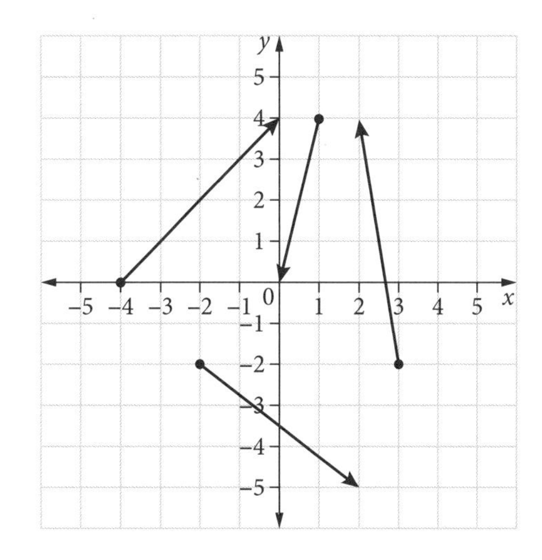

Question 3

A particle is projected with a speed of 18 m s^{-1} and an angle of projection of 45°.

Which of the following is the vector representing the particle's **initial velocity**?

A $\underset{\sim}{v} = 18\underset{\sim}{i} + 18\underset{\sim}{j}$

B $\underset{\sim}{v} = 9\sqrt{2}\underset{\sim}{i} + 9\sqrt{2}\underset{\sim}{j}$

C $\underset{\sim}{v} = 9\underset{\sim}{i} + 9\underset{\sim}{j}$

D $\underset{\sim}{v} = -18\underset{\sim}{j}$

Question 4

Which vector is NOT moving in the same direction as the other three?

A $\frac{3}{2\sqrt{2}}\underset{\sim}{i} + \frac{1}{2\sqrt{2}}\underset{\sim}{j}$

B $\underset{\sim}{i} - 3\underset{\sim}{j}$

C $\frac{3}{5}\underset{\sim}{i} + \frac{1}{5}\underset{\sim}{j}$

D $3\underset{\sim}{i} + \underset{\sim}{j}$

Question 5

A particle moves through the air according to the displacement vector $\underset{\sim}{r} = \begin{pmatrix} 4t \\ 5t - 5t^2 \end{pmatrix}$, where $\underset{\sim}{r}$ is measured in metres.

What horizontal distance will the particle travel before it hits the ground?

A 1.25 m

B 2 m

C 4 m

D 8 m

Question 6

For what values are the following vectors $\underset{\sim}{c} = 2\underset{\sim}{i} + n\underset{\sim}{j}$ and $\underset{\sim}{d} = n^2\underset{\sim}{i} - \underset{\sim}{j}$ perpendicular?

A $n = \frac{1}{2}$ and $n = 0$

B $n = -\frac{1}{2}$ and $n = 0$

C $n = 2$ and $n = 0$

D $n = -2$ and $n = 0$

Question 7

Which expression gives the angle between the vectors $\begin{pmatrix} 2 \\ 1 \end{pmatrix}$ and $\begin{pmatrix} -4 \\ 2 \end{pmatrix}$?

A $\cos^{-1}(-0.6)$ **B** $\cos^{-1}(0.6)$ **C** $\cos^{-1}(-0.06)$ **D** $\cos^{-1}(0.06)$

Question 8

A particle moves through the air according to the displacement vector $\underset{\sim}{r}(t) = \begin{pmatrix} 25t \\ 48t - 5t^2 \end{pmatrix}$.

Which of the following is the Cartesian equation of this motion?

A $y = \frac{1}{125}(240x - x^2)$ **B** $y = 48x - 5x^2$ **C** $y = \frac{1}{5}(240 - x^2)$ **D** $y = 1200t^2 - 125t^3$

Question 9

The vectors $\underset{\sim}{a}$ and $\underset{\sim}{b}$ are adjacent sides of a rhombus.

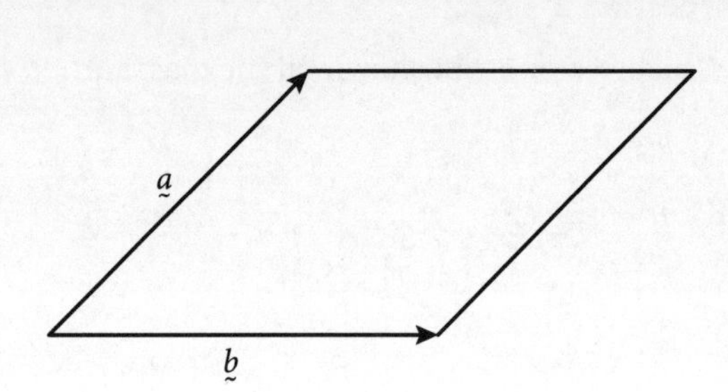

Which of the following equations are true?

A $\underset{\sim}{a} = \underset{\sim}{b}$

B $\underset{\sim}{a} \cdot \underset{\sim}{b} = 0$

C $(\underset{\sim}{a} + \underset{\sim}{b}) \cdot (\underset{\sim}{a} - \underset{\sim}{b}) = 0$

D $|\underset{\sim}{a} + \underset{\sim}{b}| = |\underset{\sim}{a} - \underset{\sim}{b}|$

Question 10

What is the graphed vector in component form?

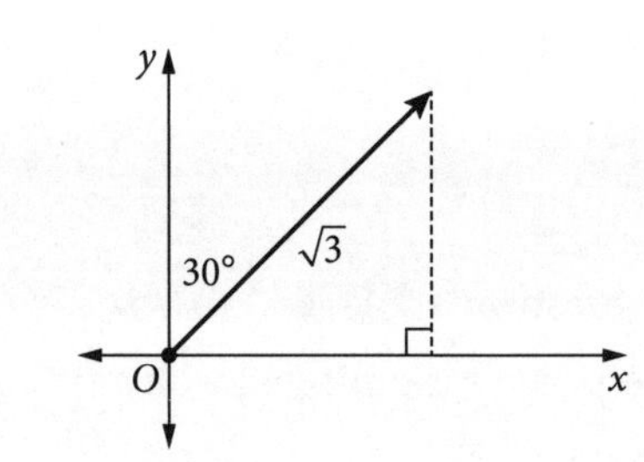

A $-\frac{\sqrt{3}}{2}\underset{\sim}{i} - \frac{3}{2}\underset{\sim}{j}$ **B** $\sqrt{3}\underset{\sim}{i} - \underset{\sim}{j}$

C $\underset{\sim}{i} + \sqrt{3}\underset{\sim}{j}$ **D** $\frac{\sqrt{3}}{2}\underset{\sim}{i} + \frac{3}{2}\underset{\sim}{j}$

Question 11

Which vector below is parallel to $\overrightarrow{OE} = -8\underset{\sim}{i} - 12\underset{\sim}{j}$?

A $\overrightarrow{OA} = 4\underset{\sim}{i} + 6\underset{\sim}{j}$ **B** $\overrightarrow{OB} = 6\underset{\sim}{i} - 9\underset{\sim}{j}$ **C** $\overrightarrow{OC} = -2\underset{\sim}{i} + 3\underset{\sim}{j}$ **D** $\overrightarrow{OD} = -6\underset{\sim}{i} + 9\underset{\sim}{j}$

Question 12 ©NESA 2020 HSC EXAM, QUESTION 6

The vectors $\underset{\sim}{a}$ and $\underset{\sim}{b}$ are shown.

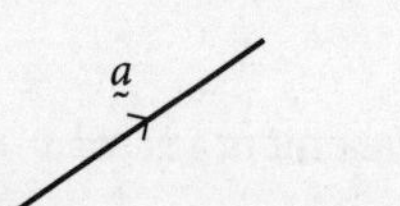

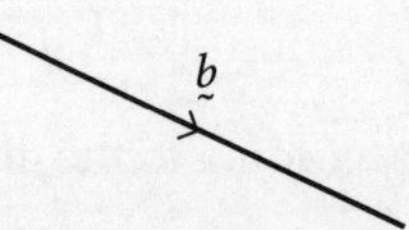

Which diagram below shows the vector $\underset{\sim}{v} = \underset{\sim}{a} - \underset{\sim}{b}$?

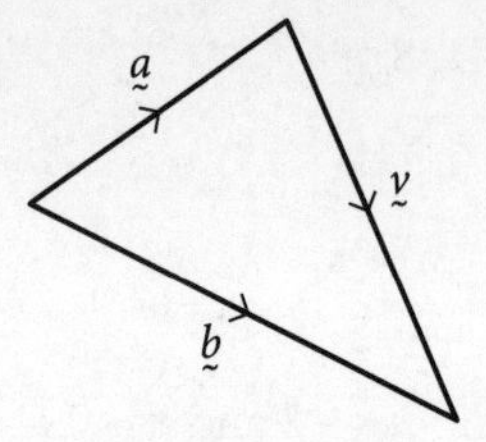

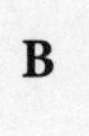
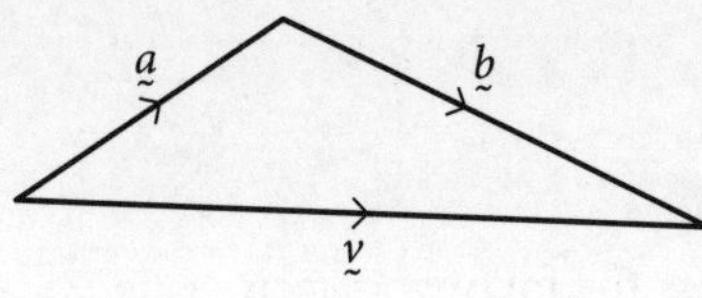

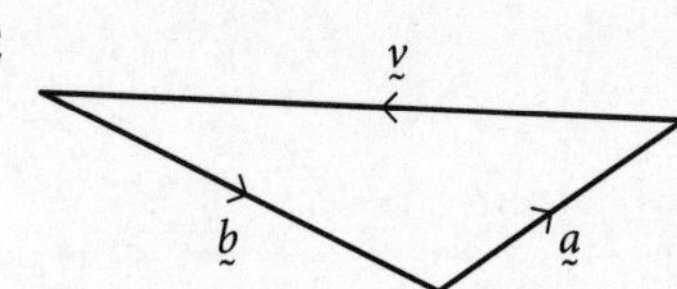

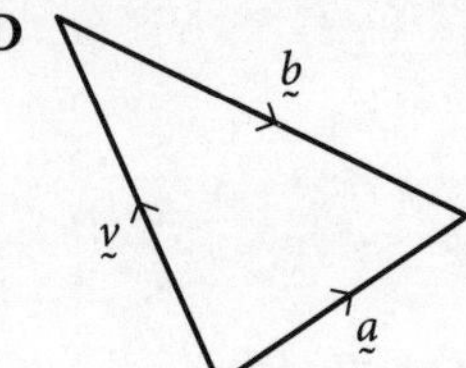

Question 13 ©NESA 2020 HSC EXAM, QUESTION 4

Maria starts at the origin and walks along all of the vector $2\underset{\sim}{i} + 3\underset{\sim}{j}$, then walks along all of the vector $3\underset{\sim}{i} - 2\underset{\sim}{j}$ and finally along all of the vector $4\underset{\sim}{i} - 3\underset{\sim}{j}$.

How far from the origin is she?

A $\sqrt{77}$

B $\sqrt{85}$

C $2\sqrt{13} + \sqrt{5}$

D $\sqrt{5} + \sqrt{7} + \sqrt{13}$

Question 14

A stone is projected horizontally from the top of a cliff 20 m above the water with a velocity of $30\,\text{m s}^{-1}$.

What horizontal distance does it travel? (Assume $g = 10\,\text{m s}^{-2}$)

A 20 m

B 30 m

C 60 m

D 120 m

Question 15

The vector $\overrightarrow{OA} = \underset{\sim}{a}$ and the vector $\overrightarrow{OH} = \underset{\sim}{h}$ and the point Q is the midpoint of F and H.

Which of the following is equivalent to the vector $\overrightarrow{EQ}$, given the grid has equally spaced lines?

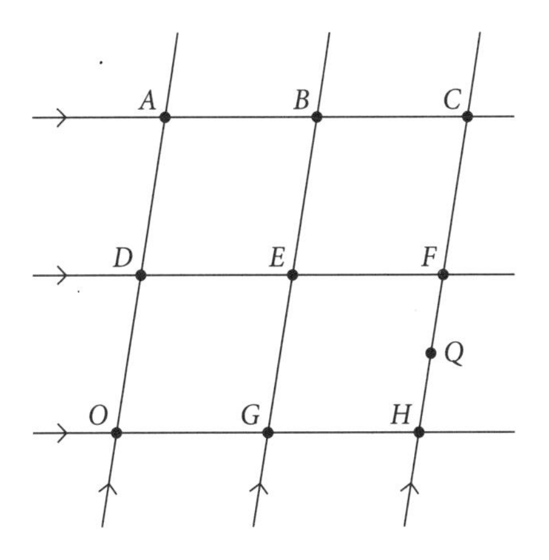

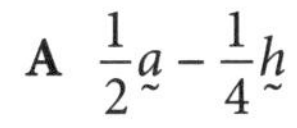

A $\frac{1}{2}\underset{\sim}{a} - \frac{1}{4}\underset{\sim}{h}$

B $\frac{1}{4}\underset{\sim}{a} - \underset{\sim}{h}$

C $-\frac{1}{4}\underset{\sim}{a} + \frac{1}{2}\underset{\sim}{h}$

D $\frac{1}{4}\underset{\sim}{a} - \frac{1}{2}\underset{\sim}{h}$

Question 16

A projectile moves through the air according to the displacement vector $\underset{\sim}{r}(t) = \begin{pmatrix} 36t \\ 18t - 5t^2 \end{pmatrix}$.

What is the initial speed and angle of projection of this projectile?

	Speed (m s^{-1})	Angle (degrees)
A	40	63
B	40	27
C	37	63
D	37	27

9780170459242

Question 17 ©NESA 2020 HSC EXAM, QUESTION 9

The projection of the vector $\begin{pmatrix}6\\7\end{pmatrix}$ onto the line $y = 2x$ is $\begin{pmatrix}4\\8\end{pmatrix}$.

The point $(6, 7)$ is reflected in the line $y = 2x$ to a point A.

What is the position vector of the point A?

A $\begin{pmatrix}6\\12\end{pmatrix}$

B $\begin{pmatrix}2\\9\end{pmatrix}$

C $\begin{pmatrix}-6\\7\end{pmatrix}$

D $\begin{pmatrix}-2\\1\end{pmatrix}$

Question 18

A stone is thrown from the top of a 15 m cliff with a velocity vector $24\underset{\sim}{i} + 10\underset{\sim}{j}$. (Assume $g = 10\ \text{m s}^{-2}$).

How much time will pass before the stone is at its **maximum height**?

A 1 second

B 2 seconds

C 3 seconds

D 4 seconds

Question 19

Find the projection of the force $\underset{\sim}{F} = c\underset{\sim}{i} + d\underset{\sim}{j}$ in the direction of the vector $m = \underset{\sim}{i} + \underset{\sim}{j}$, where c and d are constants.

A $\left(\frac{c+d}{2}\right)\underset{\sim}{m}$

B $\left(\frac{c+d}{\sqrt{2}}\right)\underset{\sim}{m}$

C $\left(\frac{c+d}{c^2+d^2}\right)\underset{\sim}{F}$

D $\frac{\underset{\sim}{F}}{c+d}$

Question 20

The equations of motion for a projection at $u\ \text{m s}^{-1}$ at an angle θ to the horizontal are $x = ut\cos\theta$ and $y = ut\sin\theta - 5t^2$. (Do NOT prove this.)

A cricketer hits a ball at $40\ \text{m s}^{-1}$ and just clears a 1 m fence 80 m away.

Which of the following is a possible angle that he hit the ball?

A 30°

B 45°

C 75°

D 80°

Practice set 2

Short-answer questions

Solutions start on page 43.

Question 1 (1 mark) 

$\underset{\sim}{a} = \underset{\sim}{i} + 2\underset{\sim}{j}$ and $\underset{\sim}{b} = -2\underset{\sim}{i} + 3\underset{\sim}{j}$.

Find $3\underset{\sim}{a} - \frac{1}{2}\underset{\sim}{b}$. 1 mark

Question 2 (2 marks) 

Use the vectors to prove that the diagonals for the rhombus *OABC* are perpendicular. 2 marks

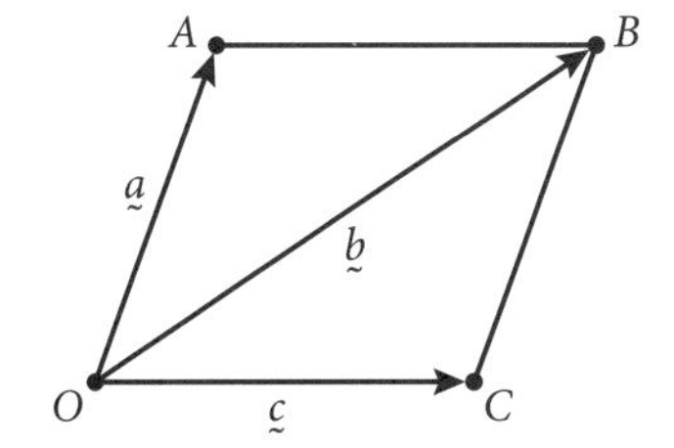

Question 3 (4 marks)

The trajectory of a projectile fired with speed $u \text{ m s}^{-1}$ at an angle θ to the horizontal is represented by the parametric equations:

$$x = ut\cos\theta \text{ and } y = ut\sin\theta - 5t^2,$$

where t is the time in seconds. (Do NOT prove this.)

a Show that the **range** of the projectile is given by $x = \dfrac{u^2 \sin 2\theta}{10}$. 2 marks

b Hence, show that the maximum range is achieved when $\theta = 45°$. 2 marks

Question 4 (3 marks)

PQRS is a parallelogram.

Given $\overrightarrow{PQ} = \underset{\sim}{a}$ and $\overrightarrow{PS} = \underset{\sim}{b}$, prove that the sum of the squares of the lengths of the diagonals is equal to the sum of the squares of the lengths of the sides. 3 marks

Question 5 (3 marks)

Consider the vector $\underset{\sim}{a} = \underset{\sim}{i} + \sqrt{3}\underset{\sim}{j}$.

a Find the unit vector in the direction of $\underset{\sim}{a}$. 1 mark

b Find the acute angle that $\underset{\sim}{a}$ makes with the positive direction of the x-axis. 1 mark

c Given that vector $\underset{\sim}{b} = m\underset{\sim}{i} - 2\underset{\sim}{j}$ is perpendicular to $\underset{\sim}{a}$, find the value of m. 1 mark

Question 6 (3 marks)

OABC is a quadrilateral.

Prove that *PQRS* is a parallelogram, given that *P*, *Q*, *R* and *S* are the midpoints of each side. 3 marks

PRACTICE SET 2

Question 7 (6 marks)

Use the vectors $\underset{\sim}{u}$, $\underset{\sim}{v}$ and $\underset{\sim}{w}$ to construct on the same axes:

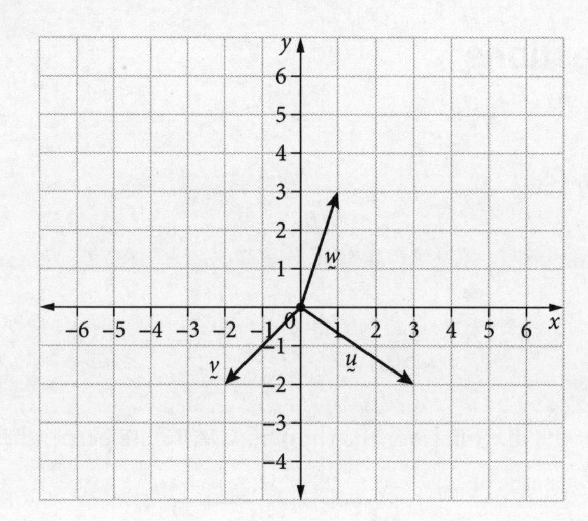

a $2\underset{\sim}{w}$ — 1 mark

b $\underset{\sim}{u} + \underset{\sim}{v}$ — 1 mark

c $-2\underset{\sim}{w} - \underset{\sim}{u}$ — 2 marks

d $\underset{\sim}{v} - \underset{\sim}{u} + 2\underset{\sim}{w}$ — 2 marks

Question 8 (2 marks)

Verify $\underset{\sim}{a} \cdot (\underset{\sim}{b} + \underset{\sim}{c}) = \underset{\sim}{a} \cdot \underset{\sim}{b} + \underset{\sim}{a} \cdot \underset{\sim}{c}$, given $\underset{\sim}{a} = 4\underset{\sim}{i} - \underset{\sim}{j}$, $\underset{\sim}{b} = 3\underset{\sim}{i} + 2\underset{\sim}{j}$ and $\underset{\sim}{c} = -2\underset{\sim}{i} + 5\underset{\sim}{j}$. — 2 marks

Question 9 (6 marks)

A ball is shot out of a cannon at $50\,\text{m s}^{-1}$ at an angle of 30°.

a Given that the acceleration vector is $\underset{\sim}{a} = \begin{pmatrix} 0 \\ -10 \end{pmatrix}$, show that $\underset{\sim}{r}(t) = \begin{pmatrix} 25\sqrt{3}t \\ -5t^2 + 25t \end{pmatrix}$. — 3 marks

b Find the Cartesian equation of the trajectory. — 1 mark

c The ball is shot towards a 25 m high cliff face that is 150 m away. — 2 marks

Determine whether the ball will hit the cliff face or fly over it.

Question 10 (2 marks)

Find the exact value of $\cos 2\theta$, given the angle between $\underset{\sim}{a} = \underset{\sim}{i} + 3\underset{\sim}{j}$ and $\underset{\sim}{b} = 3\underset{\sim}{i} + \underset{\sim}{j}$ is θ. — 2 marks

Question 11 (2 marks)

What is the magnitude of the projection of $\underset{\sim}{a} = 4\underset{\sim}{i} - 3\underset{\sim}{j}$ onto $\underset{\sim}{b} = 7\underset{\sim}{i} - \underset{\sim}{j}$, in simplest form? — 2 marks

Question 12 (5 marks)

A plane, at an altitude of 11 000 m, is travelling at a constant rate of $900\,\text{km h}^{-1}$.
It needs to drop a load of cargo to hit a target location 15 km away.

(Assume the equations of motion for a projection at $u\,\text{m s}^{-1}$ at an angle θ to the horizontal are $x = ut\cos\theta$ and $y = ut\sin\theta - 5t^2$.)

a Determine the horizontal distance required between the point where the plane drops the cargo and the location of the target. — 3 marks

b Hence, determine the amount of time required to pass before the plane releases its load of cargo. — 2 marks

Question 13 (2 marks)

LOM is the diameter of a circle with centre O and N is a point on the circumference.

Express $\overrightarrow{LN}$ in terms of $\underset{\sim}{p}$ and $\underset{\sim}{q}$ if $\underset{\sim}{p} = \overrightarrow{ON}$ and $\underset{\sim}{q} = \overrightarrow{MN}$. 2 marks

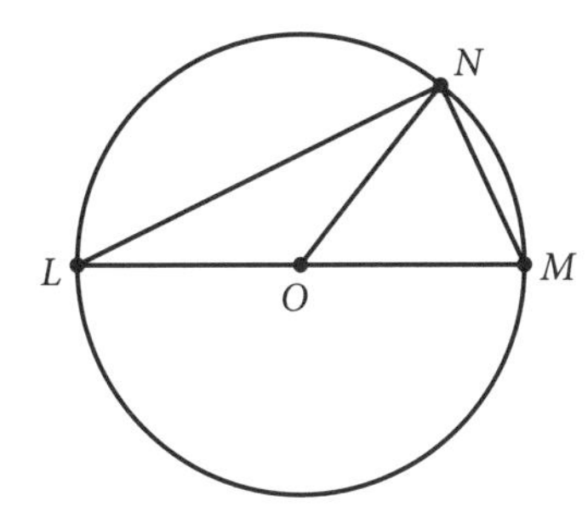

Question 14 (3 marks)

The trajectory of a projectile fired with speed $u\,\text{m s}^{-1}$ at an angle θ to the horizontal is represented by the parametric equations:

$$x = ut\cos\theta \text{ and } y = ut\sin\theta - 5t^2,$$

where t is the time in seconds. (Do NOT prove this.)

A target is shot vertically from the ground at a speed of $30\,\text{m s}^{-1}$ at a point 150 m in front of a shooter. At the same time, the shooter fires a bullet at a speed of $250\,\text{m s}^{-1}$ and hits the target mid-air.

To the nearest degree, at what angle should the shooter point his rifle to hit the target? 3 marks

Question 15 (4 marks)

a Show that the area of the triangle formed by the two vectors $\underset{\sim}{u}$, $\underset{\sim}{v}$ and the perpendicular line from $\underset{\sim}{u}$ to the tip of $\underset{\sim}{v}$ is given by 1 mark

$$A = \frac{1}{2}\left|\text{proj}_{\underset{\sim}{u}}\underset{\sim}{v}\right|\left|v - \text{proj}_{\underset{\sim}{u}}\underset{\sim}{v}\right|.$$

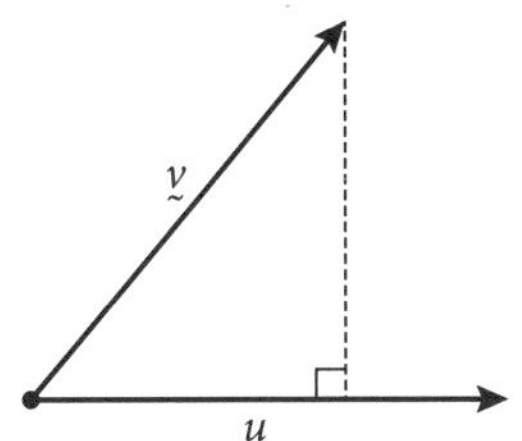

b Find the area of the triangle formed by $\underset{\sim}{u} = \begin{pmatrix}3\\4\end{pmatrix}$, $\underset{\sim}{v} = \begin{pmatrix}2\\1\end{pmatrix}$ and the orthogonal projection of $\underset{\sim}{v}$ onto $\underset{\sim}{u}$. 3 marks

Hence, find the area of the triangle formed by $\underset{\sim}{v} = \begin{pmatrix}2\\1\end{pmatrix}$, $\underset{\sim}{u} = \begin{pmatrix}3\\4\end{pmatrix}$ and the orthogonal projection of $\underset{\sim}{u}$ onto $\underset{\sim}{v}$. Give an explanation on whether the areas were equal or not.

Question 16 (3 marks)

A force $\underset{\sim}{F} = \begin{pmatrix}3\\6\end{pmatrix}$ is applied to a line $\underset{\sim}{l}$ that is parallel to the vector $\begin{pmatrix}4\\3\end{pmatrix}$.

a Find the component of the force $\underset{\sim}{F}$ in the direction of $\underset{\sim}{l}$. 2 marks

b What is the component of the force $\underset{\sim}{F}$ in the direction perpendicular to the line? 1 mark

Question 17 (5 marks)

A ball is thrown at an angle of 55° with a speed of $15\,\text{m s}^{-1}$ and hits a wall that is 10 m away, as shown in the diagram.

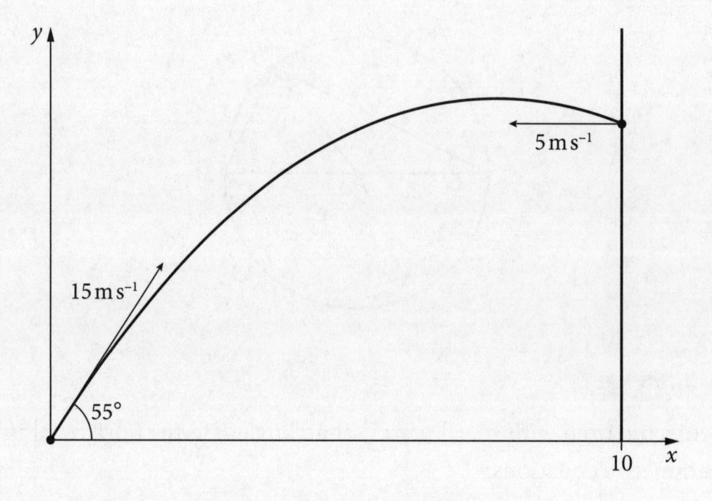

a Show that the ball hits the wall at a height of 7.5 m above the ground. 2 marks

b The ball then rebounds horizontally off the wall at a speed of $5\,\text{m s}^{-1}$. 3 marks

Determine how far away the ball is from the thrower when it hits the ground, correct to one decimal place.

Question 18 (4 marks)

A and B are points on the number plane and $\overrightarrow{AB}$ is $\begin{pmatrix} 3 \\ 1 \end{pmatrix}$.

C is another point and the area of ΔABC is 5 units^2 and $\left|\overrightarrow{AC}\right| = 4\sqrt{5}$.

Find all the possible vectors $\overrightarrow{AC}$. 4 marks

Question 19 (5 marks)

A fireworks skyrocket is constructed such that it explodes after 8 seconds. The equations of motion for the particular skyrocket are $x = 50t\cos 88°$ and $y = 50t\sin 88° - 5t^2$.

a Correct to the nearest metre, at what height does the skyrocket explode? 1 mark

b The explosion of the skyrocket triggers a second skyrocket within the first to be launched directly above the first, travelling at half the speed of the first skyrocket when it exploded. 4 marks

The second skyrocket explodes at its maximum height.

Correct to the nearest metre, at what height above the ground does this second skyrocket explode? (Assume $g = 10\,\text{m s}^{-2}$)

Question 20 (9 marks) ©NESA 2004 HSC EXAM, QUESTION 6(b) ●●●

A fire hose is at ground level on a horizontal plane. Water is projected from the hose. The angle of projection, θ, is allowed to vary. The speed of the water as it leaves the hose, v metres per second, remains constant. You may assume that if the origin is taken to be the point of projection, the path of the water is given by the parametric equations

$$x = vt\cos\theta$$

$$y = vt\sin\theta - \frac{1}{2}gt^2$$

where $g\,\text{m s}^{-2}$ is the acceleration due to **gravity.** (Do NOT prove this.)

a Show that the water returns to ground level at a distance $\dfrac{v^2\sin 2\theta}{g}$ metres from the point of projection. 2 marks

A fire hose is now aimed at a 20 m high thin wall from a point of projection at ground level 40 metres from the base of the wall. It is known that when the angle θ is 15°, the water just reaches the base of the wall.

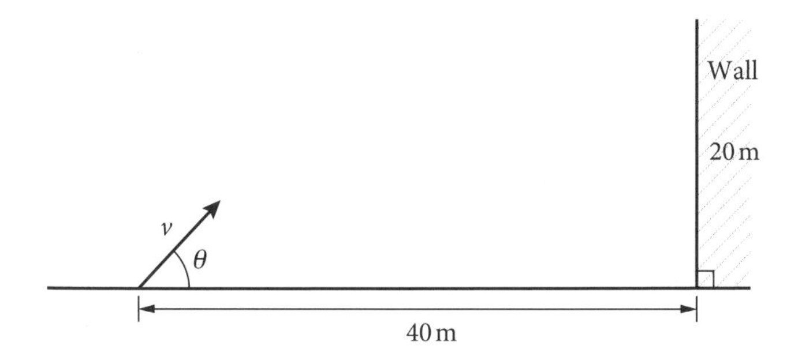

b Show that $v^2 = 80g$. 1 mark

c Show that the Cartesian equation of the path of the water is given by 2 marks

$$y = x\tan\theta - \frac{x^2\sec^2\theta}{160}.$$

d Show that the water just clears the top of the wall if $\tan^2\theta - 4\tan\theta + 3 = 0$. 2 marks

e Find all values of θ for which the water hits the front of the wall. 2 marks

Practice set 1

Worked solutions

1 B

A scalar is a quantity that has magnitude but no direction.

2 B

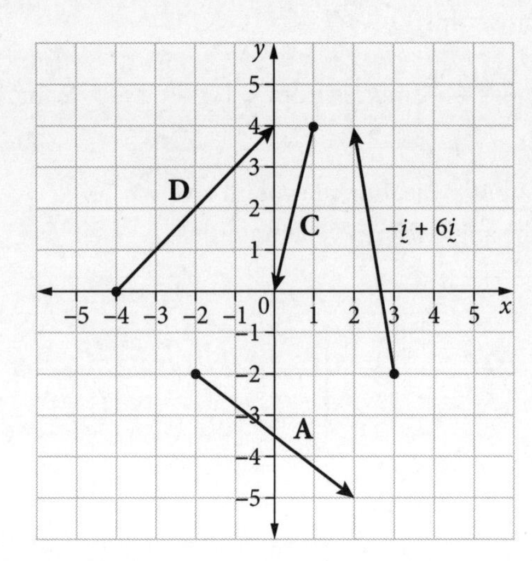

3 B

$$\dot{x} = 18\cos 45°$$
$$= 18 \times \frac{\sqrt{2}}{2}$$
$$= 9\sqrt{2}$$

$$\dot{y} = 18\sin 45°$$
$$= 18 \times \frac{\sqrt{2}}{2}$$
$$= 9\sqrt{2}$$

$$\therefore \underset{\sim}{v} = \dot{x}\underset{\sim}{i} + \dot{y}\underset{\sim}{j}$$
$$= 9\sqrt{2}\underset{\sim}{i} + 9\sqrt{2}\underset{\sim}{j}$$

4 B

The other vectors have a 'gradient' of $\frac{1}{3}$ but B has a gradient of $\frac{1}{-3}$.

5 C

$x = 4t, y = 5t - 5t^2$

Particle hits the ground when $y = 0$.

$5t(1 - t) = 0$

$t = 0, t = 1$

When $t = 1, x = 4$.

So horizontal distance = 4 m.

6 A

Perpendicular: $\underset{\sim}{a} \cdot \underset{\sim}{b} = 0$

$$\begin{pmatrix} 2 \\ n \end{pmatrix} \cdot \begin{pmatrix} n^2 \\ -1 \end{pmatrix} = 0$$
$$(2)(n^2) + (n)(-1) = 0$$
$$2n^2 - n = 0$$
$$n(2n - 1) = 0$$

$n = 0$ OR $2n - 1 = 0$

$$2n = 1$$
$$n = \frac{1}{2}$$

7 A

$$|\underset{\sim}{a}||\underset{\sim}{b}|\cos\theta = \underset{\sim}{a} \cdot \underset{\sim}{b}$$
$$\cos\theta = \frac{\underset{\sim}{a} \cdot \underset{\sim}{b}}{|\underset{\sim}{a}||\underset{\sim}{b}|}$$

$$\underset{\sim}{a} \cdot \underset{\sim}{b} = \begin{pmatrix} 2 \\ 1 \end{pmatrix} \cdot \begin{pmatrix} -4 \\ 2 \end{pmatrix}$$
$$= (2)(-4) + (1)(2)$$
$$= -6$$

$$|\underset{\sim}{a}| = \sqrt{2^2 + (-4)^2}$$
$$= \sqrt{20}$$

$$|\underset{\sim}{b}| = \sqrt{1^2 + 2^2}$$
$$= \sqrt{5}$$

$$\cos\theta = \frac{-6}{\sqrt{20}\sqrt{5}}$$
$$= \frac{-6}{10}$$
$$= -0.6$$
$$\theta = \cos^{-1}(-0.6)$$

8 A

$x = 25t \qquad y = 48t - 5t^2 \qquad [1]$

$t = \frac{x}{25} \qquad [2]$

Substitute [1] into [2]:

$$y = 48\left(\frac{x}{25}\right) - 5\left(\frac{x}{25}\right)^2$$
$$= \frac{48x}{25} - \frac{5x^2}{625}$$
$$= \frac{1}{125}(240x - x^2)$$

9 C

$\underset{\sim}{a} \neq \underset{\sim}{b}$ (not correct)

$\underset{\sim}{a} \cdot \underset{\sim}{b} = 0$ Perpendicular (not correct)

$\underset{\sim}{a} + \underset{\sim}{b} = \overrightarrow{OC}$

$\underset{\sim}{a} - \underset{\sim}{b} = \overrightarrow{BA}$

$\overrightarrow{OC} \cdot \overrightarrow{BA} = 0$ The diagonals of a rhombus are perpendicular (correct!)

10 D

$$x = |\underset{\sim}{u}|\cos\theta = \sqrt{3}\cos 60° = \sqrt{3} \times \frac{1}{2} = \frac{\sqrt{3}}{2}$$

$$y = |\underset{\sim}{u}|\sin\theta = \sqrt{3}\sin 60° = \sqrt{3} \times \frac{\sqrt{3}}{2} = \frac{3}{2}$$

So $\frac{\sqrt{3}}{2}\underset{\sim}{i} + \frac{3}{2}\underset{\sim}{j}$.

11 A

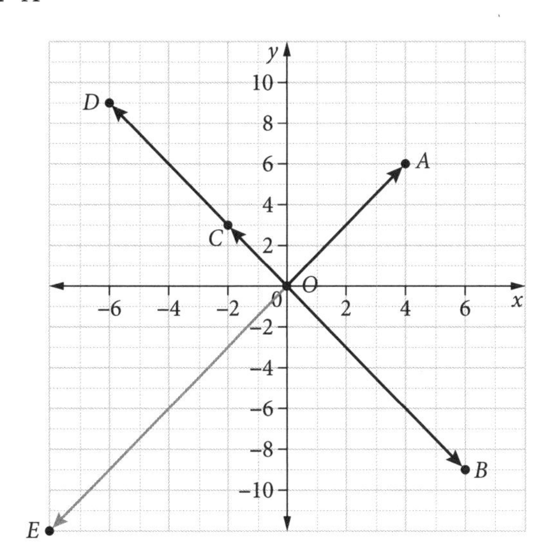

$$OE: \tan\theta = \frac{-12}{-8} = \frac{3}{2}$$

$$OA: \tan\theta = \frac{6}{4} = \frac{3}{2}$$

So vectors are parallel but in opposite directions.

12 D

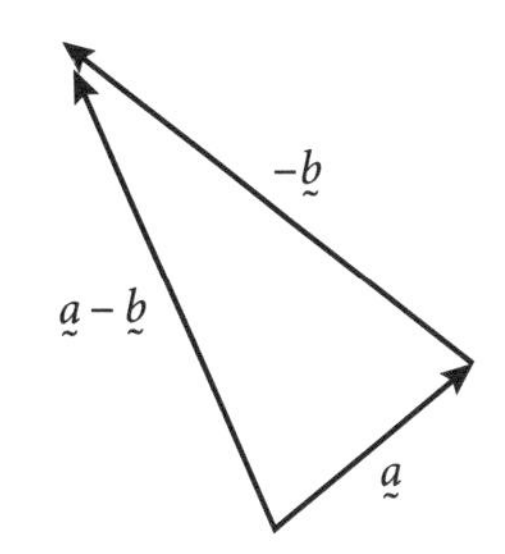

13 B

$$\underset{\sim}{v} = \begin{pmatrix} 2 \\ 3 \end{pmatrix} + \begin{pmatrix} 3 \\ -2 \end{pmatrix} + \begin{pmatrix} 4 \\ -3 \end{pmatrix} = \begin{pmatrix} 2+3+4 \\ 3+(-2)+(-3) \end{pmatrix} = \begin{pmatrix} 9 \\ -2 \end{pmatrix}$$

$$|\underset{\sim}{v}| = \sqrt{9^2 + (-2)^2} = \sqrt{85}$$

14 C

Determine equations of motion.

$\ddot{x} = 0$	$\ddot{y} = -10$
$\dot{x} = c_1$	$\dot{y} = -10t + c_3$
$t = 0, \dot{x} = 30$	$t = 0, \dot{y} = 0$
$c_1 = 30$	$c_3 = 0$
$\therefore \dot{x} = 30$	$\therefore \dot{y} = -10t$
$x = 30t + c_2$	$y = -5t^2 + c_4$
$t = 0, x = 0$	$t = 0, y = 20$
$c_2 = 0$	$c_4 = 20$
$\therefore x = 30t$	$\therefore y = 20 - 5t^2$

When $y = 0$:

$$20 - 5t^2 = 0$$
$$5t^2 = 20$$
$$t^2 = 4$$
$$t = \pm 2$$
$$\therefore t = 2$$

Substitute into $x = 30t$

$$x = 30 \times 2 = 60\,\text{m}$$

15 C

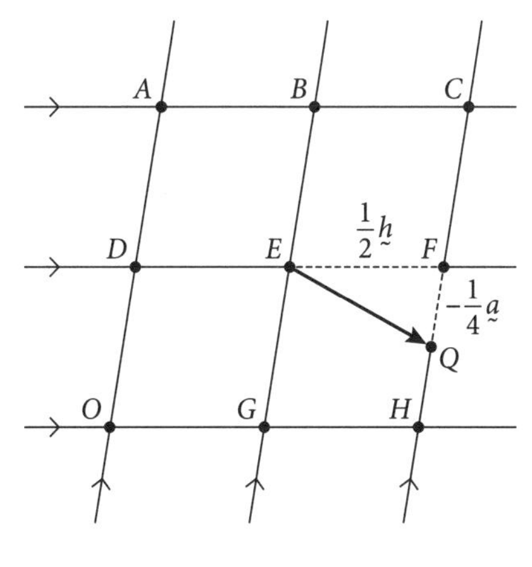

$$\overrightarrow{EQ} = \frac{1}{2}\underset{\sim}{h} - \frac{1}{4}\underset{\sim}{a} = -\frac{1}{4}\underset{\sim}{a} + \frac{1}{2}\underset{\sim}{h}$$

16 B

$$\underset{\sim}{r}(t) = \begin{pmatrix} 36t \\ 18t - 5t^2 \end{pmatrix}$$

$$\underset{\sim}{v}(t) = \begin{pmatrix} 36 \\ 18 - 10t \end{pmatrix}$$

$$\underset{\sim}{v}(0) = \begin{pmatrix} 36 \\ 18 \end{pmatrix}$$

$$\begin{aligned} |\underset{\sim}{v}(0)| &= \sqrt{36^2 + 18^2} \\ &= \sqrt{1620} \\ &= 40.25 \\ &\approx 40\,\text{m}\,\text{s}^{-1} \text{ (nearest m s}^{-1}\text{)} \end{aligned}$$

$$\begin{aligned} \tan\theta &= \frac{\dot{y}}{\dot{x}} \\ &= \frac{18}{36} \\ \theta &= \tan^{-1}\left(\frac{18}{36}\right) \\ &= 26°34' \\ &\approx 27° \text{ (nearest degree)} \end{aligned}$$

17 B

Graph the information and the reflection:

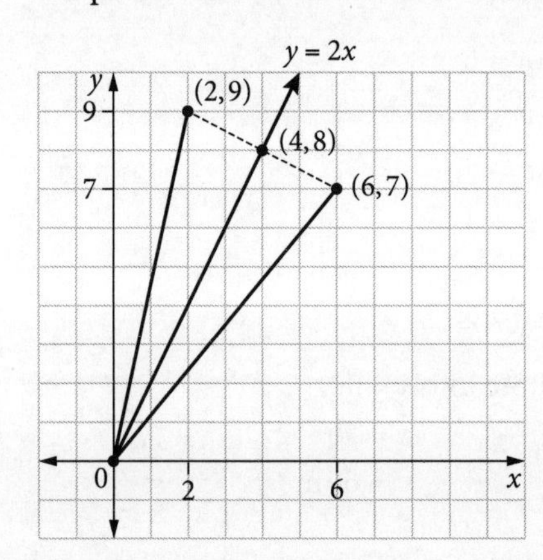

18 A

Determine equations of motion.

$\ddot{x} = 0$	$\ddot{y} = -10$
$\dot{x} = c_1$	$\dot{y} = -10t + c_2$
$t = 0, \dot{x} = 24$	$t = 0, \dot{y} = 10$
$c_1 = 24$	$c_2 = 10$
$\therefore \dot{x} = 24$	$\therefore \dot{y} = 10 - 10t$

Maximum height when $\dot{y} = 0$:

$$\begin{aligned} 10 - 10t &= 0 \\ 10t &= 10 \\ t &= 1 \text{ second} \end{aligned}$$

19 A

$\text{proj}_{\underset{\sim}{u}}\underset{\sim}{v} = \dfrac{\underset{\sim}{u}\cdot\underset{\sim}{v}}{|\underset{\sim}{u}|^2}\underset{\sim}{u}$ (projection of $\underset{\sim}{v}$ onto $\underset{\sim}{u}$)

$\text{proj}_{\underset{\sim}{m}}\underset{\sim}{F} = \dfrac{\underset{\sim}{m}\cdot\underset{\sim}{F}}{|\underset{\sim}{m}|^2}\underset{\sim}{m}$ (projection of $\underset{\sim}{F}$ onto $\underset{\sim}{m}$)

Step 1: $\underset{\sim}{u}\cdot\underset{\sim}{v} = x_1x_2 + y_1y_2$

$$\begin{aligned} \underset{\sim}{m}\cdot\underset{\sim}{F} &= (1)(c) + (1)(d) \\ &= c + d \end{aligned}$$

Step 2: $|\underset{\sim}{a}| = \sqrt{x^2 + y^2}$

$$\begin{aligned} |\underset{\sim}{m}| &= \sqrt{1^2 + 1^2} \\ &= \sqrt{2} \end{aligned}$$

Step 3: $\text{proj}_{\underset{\sim}{m}}\underset{\sim}{F} = \dfrac{\underset{\sim}{m}\cdot\underset{\sim}{F}}{|\underset{\sim}{m}|^2}\underset{\sim}{m}$

$$\begin{aligned} &= \frac{c + d}{(\sqrt{2})^2}\underset{\sim}{m} \\ &= \left(\frac{c + d}{2}\right)\underset{\sim}{m} \end{aligned}$$

20 C

$x = ut\cos\theta$	$y = ut\sin\theta - 5t^2$
$= 40t\cos\theta$	$= 40t\sin\theta - 5t^2$

When $x = 80$, $y = 1$:

$80 = 40t\cos\theta$ $\qquad$ $1 = 40t\sin\theta - 5t^2$ [2]

$t = \dfrac{2}{\cos\theta}$ [1]

Substitute [1] into [2]:

$$\begin{aligned} 1 &= \frac{80\sin\theta}{\cos\theta} - 5\left(\frac{2}{\cos\theta}\right)^2 \\ &= 80\tan\theta - \frac{20}{\cos^2\theta} \\ &= 80\tan\theta - 20\sec^2\theta \\ &= 80\tan\theta - 20\tan^2\theta - 20 \end{aligned}$$

$$20\tan^2\theta - 80\tan\theta + 21 = 0$$

$$\begin{aligned} \tan\theta &= \frac{80 \pm \sqrt{80^2 - 4(20)(21)}}{2 \times 20} \\ &= 0.28244, 3.7176 \\ \theta &= 15°46', 74°56' \\ &\approx 16°, 75° \end{aligned}$$

Practice set 2

Worked solutions

Question 1

$$3\underset{\sim}{a} - \frac{1}{2}\underset{\sim}{b} = 3\begin{pmatrix}1\\2\end{pmatrix} - \frac{1}{2}\begin{pmatrix}-2\\3\end{pmatrix}$$
$$= \begin{pmatrix}3\\6\end{pmatrix} - \begin{pmatrix}-1\\\frac{3}{2}\end{pmatrix}$$
$$= \begin{pmatrix}4\\\frac{9}{2}\end{pmatrix}$$

So $3\underset{\sim}{a} - \frac{1}{2}\underset{\sim}{b} = 4\underset{\sim}{i} + \frac{9}{2}\underset{\sim}{j}$.

Question 2

$|\underset{\sim}{a}| = |\underset{\sim}{c}|$ ($OABC$ is a rhombus)

$\overrightarrow{OB} = \underset{\sim}{a} + \underset{\sim}{c}$

$\overrightarrow{AC} = \underset{\sim}{c} - \underset{\sim}{a}$

$$(\underset{\sim}{c} - \underset{\sim}{a}) \cdot (\underset{\sim}{a} + \underset{\sim}{c}) = \underset{\sim}{c}(\underset{\sim}{a} + \underset{\sim}{c}) - \underset{\sim}{a}(\underset{\sim}{a} + \underset{\sim}{c})$$
$$= \underset{\sim}{c} \cdot \underset{\sim}{a} + \underset{\sim}{c} \cdot \underset{\sim}{c} - \underset{\sim}{a} \cdot \underset{\sim}{a} - \underset{\sim}{a} \cdot \underset{\sim}{c}$$
$$= |\underset{\sim}{c}|^2 - |\underset{\sim}{a}|^2$$
$$= 0$$

So $\overrightarrow{OB} \perp \overrightarrow{AC}$.

Question 3

a Projectile reaches range when it returns to the ground.

When $y = 0$:

$$ut\sin\theta - 5t^2 = 0$$
$$t(u\sin\theta - 5t) = 0$$
$$t = 0,\ t = \frac{u\sin\theta}{5}$$

When $t = \frac{u\sin\theta}{5}$:

$$x = u\cos\theta \times \frac{u\sin\theta}{5}$$
$$= \frac{u^2\sin\theta\cos\theta}{5}$$
$$= \frac{2u^2\sin\theta\cos\theta}{10}$$
$$= \frac{u^2\sin 2\theta}{10} \qquad (2\sin\theta\cos\theta = \sin 2\theta)$$

b $\frac{dx}{d\theta} = \frac{2u^2\cos 2\theta}{10}$

When $\frac{dx}{d\theta} = 0$:

$$\frac{2u^2\cos 2\theta}{10} = 0$$
$$\cos 2\theta = 0$$
$$2\theta = 90°, 270°, \ldots$$
$$\theta = 45°, 135°, \ldots$$

$0 < \theta < 90°$

The maximum range is reached when $\theta = 45°$.

WORKED SOLUTIONS

Question 4

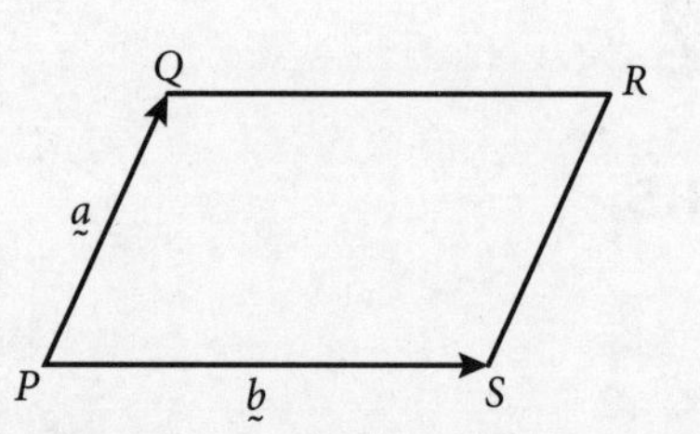

$\overrightarrow{PR} = \underset{\sim}{a} + \underset{\sim}{b}$

$\overrightarrow{SQ} = \underset{\sim}{a} - \underset{\sim}{b}$

$\overrightarrow{PQ} = \overrightarrow{SR} = \underset{\sim}{a}$ ($PQRS$ is a parallelogram)

$\overrightarrow{PS} = \overrightarrow{QR} = \underset{\sim}{b}$

Proving:

$$\left|\overrightarrow{PR}\right|^2 + \left|\overrightarrow{SQ}\right|^2 = 2\left|\overrightarrow{PQ}\right|^2 + 2\left|\overrightarrow{PS}\right|^2$$

$$|\underset{\sim}{a} + \underset{\sim}{b}|^2 + |\underset{\sim}{a} - \underset{\sim}{b}|^2 = 2|\underset{\sim}{a}|^2 + 2|\underset{\sim}{b}|^2$$

$$\begin{aligned}|\underset{\sim}{a} + \underset{\sim}{b}|^2 + |\underset{\sim}{a} - \underset{\sim}{b}|^2 &= (\underset{\sim}{a} + \underset{\sim}{b}) \cdot (\underset{\sim}{a} + \underset{\sim}{b}) + (\underset{\sim}{a} - \underset{\sim}{b}) \cdot (\underset{\sim}{a} - \underset{\sim}{b}) \\ &= (\underset{\sim}{a} \cdot \underset{\sim}{a} + \underset{\sim}{a} \cdot \underset{\sim}{b} + \underset{\sim}{b} \cdot \underset{\sim}{a} + \underset{\sim}{b} \cdot \underset{\sim}{b}) + (\underset{\sim}{a} \cdot \underset{\sim}{a} + \underset{\sim}{a} \cdot -\underset{\sim}{b} - \underset{\sim}{b} \cdot \underset{\sim}{a} - \underset{\sim}{b} \cdot -\underset{\sim}{b}) \\ &= (|\underset{\sim}{a}|^2 + 2\underset{\sim}{a} \cdot \underset{\sim}{b} + |\underset{\sim}{b}|^2) + (|\underset{\sim}{a}|^2 - 2\underset{\sim}{a} \cdot \underset{\sim}{b} + |\underset{\sim}{b}|^2) \\ &= |\underset{\sim}{a}|^2 + 2\underset{\sim}{a} \cdot \underset{\sim}{b} + |\underset{\sim}{b}|^2 + |\underset{\sim}{a}|^2 - 2\underset{\sim}{a} \cdot \underset{\sim}{b} + |\underset{\sim}{b}|^2 \\ &= 2|\underset{\sim}{a}|^2 + 2|\underset{\sim}{b}|^2\end{aligned}$$

Question 5

a $$\begin{aligned}\hat{\underset{\sim}{a}} &= \frac{\underset{\sim}{a}}{|\underset{\sim}{a}|} \\ &= \frac{\underset{\sim}{i} + \sqrt{3}\underset{\sim}{j}}{\sqrt{(1)^2 + (\sqrt{3})^2}} \\ &= \frac{\underset{\sim}{i} + \sqrt{3}\underset{\sim}{j}}{2} \\ &= \frac{1}{2}(\underset{\sim}{i} + \sqrt{3}\underset{\sim}{j})\end{aligned}$$

b $$\begin{aligned}\tan\theta &= \frac{y}{x} \\ \tan\theta &= \frac{\sqrt{3}}{1} \\ \theta &= \tan^{-1}\sqrt{3} \\ \theta &= 60°\end{aligned}$$

c Clue: perpendicular $\underset{\sim}{a} \cdot \underset{\sim}{b} = 0$

$$\begin{aligned}\underset{\sim}{a} \cdot \underset{\sim}{b} &= 0 \\ (\underset{\sim}{i} + \sqrt{3}\underset{\sim}{j}) \cdot (m\underset{\sim}{i} - 2\underset{\sim}{j}) &= 0 \\ \begin{pmatrix}1 \\ \sqrt{3}\end{pmatrix} \cdot \begin{pmatrix}m \\ -2\end{pmatrix} &= 0 \\ m - 2\sqrt{3} &= 0 \\ m &= 2\sqrt{3}\end{aligned}$$

Question 6

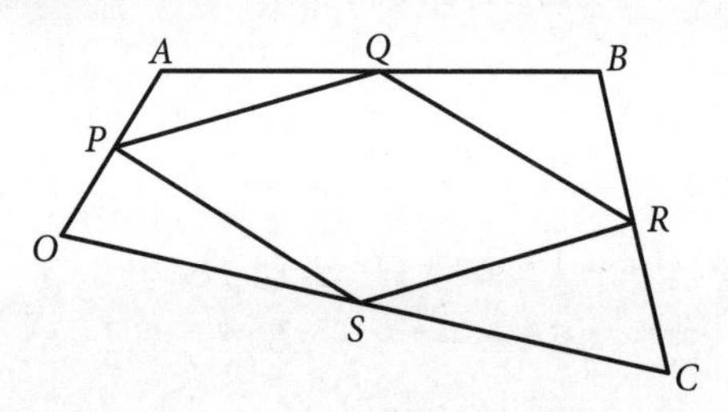

Proving opposite sides have equal length:
$\overrightarrow{PQ} = \overrightarrow{SR}$ and $\overrightarrow{QR} = \overrightarrow{PS}$

$$\begin{aligned}\overrightarrow{OB} &= \overrightarrow{OA} + \overrightarrow{AB} \\ &= \overrightarrow{OC} + \overrightarrow{CB} \\ \overrightarrow{PQ} &= \overrightarrow{PA} + \overrightarrow{AQ} \\ &= \frac{1}{2}(\overrightarrow{OA} + \overrightarrow{AB}) \\ &= \frac{1}{2}\overrightarrow{OB} \\ \overrightarrow{SR} &= \overrightarrow{SC} + \overrightarrow{CR} \\ &= \frac{1}{2}(\overrightarrow{OC} + \overrightarrow{CB}) \\ &= \frac{1}{2}\overrightarrow{OB}\end{aligned}$$

$\therefore \overrightarrow{PQ} = \overrightarrow{SR}$.

$$\begin{aligned}\overrightarrow{AC} &= \overrightarrow{AB} + \overrightarrow{BC} \\ &= \overrightarrow{AO} + \overrightarrow{OC} \\ \overrightarrow{QR} &= \overrightarrow{QB} + \overrightarrow{BR} \\ &= \frac{1}{2}(\overrightarrow{AB} + \overrightarrow{BC}) \\ &= \frac{1}{2}\overrightarrow{AC} \\ \overrightarrow{PS} &= \overrightarrow{PO} + \overrightarrow{OS} \\ &= \frac{1}{2}(\overrightarrow{AO} + \overrightarrow{OC}) \\ &= \frac{1}{2}\overrightarrow{AC}\end{aligned}$$

$\therefore \overrightarrow{QR} = \overrightarrow{PS}$.

So $PQRS$ is a parallelogram.

Question 7

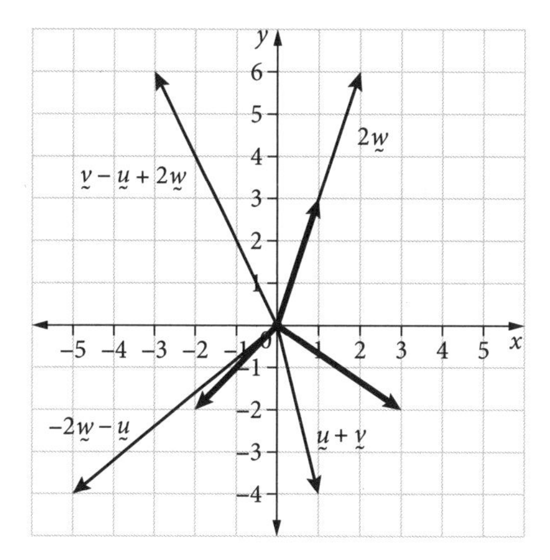

Question 8

$$\underset{\sim}{a} \cdot (\underset{\sim}{b} + \underset{\sim}{c}) = \begin{pmatrix} 4 \\ -1 \end{pmatrix} \cdot \left(\begin{pmatrix} 3 \\ 2 \end{pmatrix} + \begin{pmatrix} -2 \\ 5 \end{pmatrix} \right)$$

$$= \begin{pmatrix} 4 \\ -1 \end{pmatrix} \cdot \begin{pmatrix} 1 \\ 7 \end{pmatrix}$$

$$= (4)(1) + (-1)(7)$$

$$= -3$$

$$\underset{\sim}{a} \cdot \underset{\sim}{b} + \underset{\sim}{a} \cdot \underset{\sim}{c}$$

$$= \begin{pmatrix} 4 \\ -1 \end{pmatrix} \cdot \begin{pmatrix} 3 \\ 2 \end{pmatrix} + \begin{pmatrix} 4 \\ -1 \end{pmatrix} \cdot \begin{pmatrix} -2 \\ 5 \end{pmatrix}$$

$$= (4)(3) + (-1)(2) + (4)(-2) + (-1)(5)$$

$$= 10 + (-13)$$

$$= -3$$

So $\underset{\sim}{a} \cdot (\underset{\sim}{b} + \underset{\sim}{c}) = \underset{\sim}{a} \cdot \underset{\sim}{b} + \underset{\sim}{a} \cdot \underset{\sim}{c}$.

Question 9

a Determine initial conditions:

When $t = 0$:

$$\underset{\sim}{v} = \begin{pmatrix} \underset{\sim}{u}\cos\theta \\ \underset{\sim}{u}\sin\theta \end{pmatrix} = \begin{pmatrix} 50\cos 30° \\ 50\sin 30° \end{pmatrix}$$

$$= \begin{pmatrix} 25\sqrt{3} \\ 25 \end{pmatrix}$$

$$\underset{\sim}{r} = \begin{pmatrix} 0 \\ 0 \end{pmatrix}$$

$$\underset{\sim}{a} = \begin{pmatrix} 0 \\ -10 \end{pmatrix}$$

$$\underset{\sim}{v} = \int \begin{pmatrix} 0 \\ -10 \end{pmatrix} dt$$

$$= \begin{pmatrix} 0 \\ -10 \end{pmatrix} t + \underset{\sim}{c}_1$$

When $t = 0$:

$$\underset{\sim}{v} = \begin{pmatrix} 25\sqrt{3} \\ 25 \end{pmatrix}$$

$$\underset{\sim}{c}_1 = \begin{pmatrix} 25\sqrt{3} \\ 25 \end{pmatrix}$$

So $\underset{\sim}{v} = \begin{pmatrix} 25\sqrt{3} \\ 25 - 10t \end{pmatrix}$.

$$\underset{\sim}{r} = \int \begin{pmatrix} 25\sqrt{3} \\ 25 - 10t \end{pmatrix} dt$$

$$= \begin{pmatrix} 25\sqrt{3}t \\ 25t - 5t^2 \end{pmatrix} + \underset{\sim}{c}_2$$

When $t = 0$:

$$\underset{\sim}{r} = \begin{pmatrix} 0 \\ 0 \end{pmatrix}$$

$$\underset{\sim}{c}_2 = \begin{pmatrix} 0 \\ 0 \end{pmatrix}$$

So $\underset{\sim}{r} = \begin{pmatrix} 25\sqrt{3}t \\ 25t - 5t^2 \end{pmatrix}$.

b Using part **a**:

$x = 25t\sqrt{3} \quad y = 25t - 5t^2 \quad [2]$

$t = \dfrac{x}{25\sqrt{3}} \quad [1]$

Substitute [1] into [2]:

$$\begin{aligned} y &= 25\left(\frac{x}{25\sqrt{3}}\right) - 5\left(\frac{x}{25\sqrt{3}}\right)^2 \\ &= \frac{x}{\sqrt{3}} - \frac{5x^2}{1875} \\ &= \frac{x\sqrt{3}}{3} - \frac{x^2}{375} \\ &= \frac{1}{375}(125x\sqrt{3} - x^2) \end{aligned}$$

c When $x = 150$:

$$\begin{aligned} y &= \frac{1}{375}(125 \times 150\sqrt{3} - 150^2) \\ &= 26.6\,\text{m} \end{aligned}$$

The ball will fly over the cliff face.

Question 10

$\cos 2\theta = 2\cos^2\theta - 1$ (from trigonometric equations topic, see Chapter 3)

Step 1: $|\underset{\sim}{a}||\underset{\sim}{b}|\cos\theta = \underset{\sim}{a} \cdot \underset{\sim}{b}$

$$\cos\theta = \frac{\underset{\sim}{a} \cdot \underset{\sim}{b}}{|\underset{\sim}{a}||\underset{\sim}{b}|}$$

$$\begin{aligned} \underset{\sim}{a} \cdot \underset{\sim}{b} &= \begin{pmatrix}1\\3\end{pmatrix} \cdot \begin{pmatrix}3\\1\end{pmatrix} \\ &= (1)(3) + (3)(1) \\ &= 6 \end{aligned}$$

$$\begin{aligned} |\underset{\sim}{a}| &= \sqrt{1^2 + 3^2} \\ &= \sqrt{10} \end{aligned}$$

$$\begin{aligned} |\underset{\sim}{b}| &= \sqrt{3^2 + 1^2} \\ &= \sqrt{10} \end{aligned}$$

$$\begin{aligned} \cos\theta &= \frac{6}{\sqrt{10}\sqrt{10}} \\ &= \frac{3}{5} \end{aligned}$$

Step 2:
$$\begin{aligned} \cos 2\theta &= 2\cos^2\theta - 1 \\ &= 2\left(\frac{3}{5}\right)^2 - 1 \\ &= -\frac{7}{25} \end{aligned}$$

Question 11

$$\begin{aligned} \left|\text{proj}_{\underset{\sim}{b}}\underset{\sim}{a}\right| &= \frac{\underset{\sim}{b} \cdot \underset{\sim}{a}}{|\underset{\sim}{b}|} \\ &= \frac{31}{\sqrt{50}} \\ &= \frac{31}{\sqrt{25}\sqrt{2}} \\ &= \frac{31}{5\sqrt{2}} \\ &= \frac{31}{5\sqrt{2}} \times \frac{\sqrt{2}}{\sqrt{2}} \end{aligned}$$

(rationalise the denominator)

$$= \frac{31\sqrt{2}}{10}$$

$$\begin{aligned} \underset{\sim}{b} \cdot \underset{\sim}{a} &= \begin{pmatrix}7\\-1\end{pmatrix} \cdot \begin{pmatrix}4\\-3\end{pmatrix} \\ &= (7)(4) + (-1)(-3) \\ &= 31 \end{aligned}$$

$$\begin{aligned} |\underset{\sim}{b}| &= \sqrt{7^2 + (-1)^2} \\ &= \sqrt{50} \end{aligned}$$

Question 12

a Note: $900\,\text{km h}^{-1} = 250\,\text{m s}^{-1}$

Initial conditions: when $t = 0$:

$\dot{x} = 250, \quad \dot{y} = 0$

$x = 0, \quad y = 11\,000$

Determine equations of motion.

$\ddot{x} = 0$	$\ddot{y} = -10$
$\dot{x} = c_1$	$\dot{y} = -10t + c_3$
$t = 0, \dot{x} = 250$	$t = 0, \dot{y} = 0$
$c_1 = 250$	$c_3 = 0$
$\therefore \dot{x} = 250$	$\therefore \dot{y} = -10t$
$x = 250t + c_2$	$y = -5t^2 + c_4$
$t = 0, x = 0$	$t = 0, y = 11\,000$
$c_2 = 0$	$c_4 = 11\,000$
$\therefore x = 250t$	$\therefore y = 11\,000 - 5t^2$

When $y = 0$:

$$\begin{aligned} 0 &= 11\,000 - 5t^2 \\ 5t^2 &= 11\,000 \\ t^2 &= 2200 \\ t &\approx 46.9 \ (t > 0) \\ &\approx 47 \text{ seconds} \end{aligned}$$

Then
$$\begin{aligned} x &= 250t \\ &= 250 \times 47 \\ &= 11\,750\,\text{m.} \end{aligned}$$

So 11.75 km distance is required between the drop point and the target.

b Remaining distance required $= 15 - 11.75 = 3.25\,\text{km}$

$$\text{Time required} = \frac{3.25\text{km}}{900\text{km/h}} = 0.003\,611\ldots \text{ hours} = 13 \text{ seconds}$$

Question 13

$$\overrightarrow{LN} = \overrightarrow{LM} + \overrightarrow{MN} = \overrightarrow{LM} + \underset{\sim}{q}$$

$$\overrightarrow{LN} = \overrightarrow{LO} + \overrightarrow{ON} = \frac{1}{2}\overrightarrow{LM} + \overrightarrow{ON} = \frac{1}{2}\overrightarrow{LM} + \underset{\sim}{p}$$

$$\therefore \overrightarrow{LM} + \underset{\sim}{q} = \frac{1}{2}\overrightarrow{LM} + \underset{\sim}{p}$$

$$\frac{1}{2}\overrightarrow{LM} + \underset{\sim}{q} = \underset{\sim}{p}$$

$$\frac{1}{2}\overrightarrow{LM} = \underset{\sim}{p} - \underset{\sim}{q}$$

$$\overrightarrow{LM} = 2\underset{\sim}{p} - 2\underset{\sim}{q}$$

$$\therefore \overrightarrow{LN} = \overrightarrow{LM} + \overrightarrow{MN} = (2\underset{\sim}{p} - 2\underset{\sim}{q}) + \underset{\sim}{q} = 2\underset{\sim}{p} - \underset{\sim}{q}$$

Question 14

Determine equations of motion:

Bullet:

$x_b = 250t\cos\theta,\ y_b = 250t\sin\theta - 5t^2$

Target:

$x_t = 150,\ y_t = 30t - 5t^2$

If bullet and target collide, $y_b = y_t$.

$$250t\sin\theta - 5t^2 = 30t - 5t^2$$
$$250t\sin\theta = 30t$$
$$\sin\theta = \frac{30}{250}$$
$$\theta = \sin^{-1}\left(\frac{30}{250}\right) \approx 6.8921° \approx 7°$$

Question 15

a Area of a triangle $= \dfrac{bh}{2}$

Area of a triangle $= \dfrac{|\text{proj}_{\underset{\sim}{u}}\underset{\sim}{v}||\underset{\sim}{v} - \text{proj}_{\underset{\sim}{u}}\underset{\sim}{v}|}{2}$

(the orthogonal projection is the height)

So $A = \frac{1}{2}|\text{proj}_{\underset{\sim}{u}}\underset{\sim}{v}||\underset{\sim}{v} - \text{proj}_{\underset{\sim}{u}}\underset{\sim}{v}|$.

b Area of 1st triangle: $A = \frac{1}{2}|\text{proj}_{\underset{\sim}{u}}\underset{\sim}{v}||\underset{\sim}{v} - \text{proj}_{\underset{\sim}{u}}\underset{\sim}{v}|$

$$|\text{proj}_{\underset{\sim}{u}}\underset{\sim}{v}| = \frac{\underset{\sim}{u}\cdot\underset{\sim}{v}}{|\underset{\sim}{u}|} = \frac{10}{5} = 2$$

$$\underset{\sim}{u}\cdot\underset{\sim}{v} = \begin{pmatrix}3\\4\end{pmatrix}\cdot\begin{pmatrix}2\\1\end{pmatrix} = (3)(2) + (4)(1) = 10$$

$$|\underset{\sim}{u}| = \sqrt{3^2 + 4^2} = 5$$

$$|\underset{\sim}{v} - \text{proj}_{\underset{\sim}{u}}\underset{\sim}{v}| = \left|\underset{\sim}{v} - \frac{\underset{\sim}{u}\cdot\underset{\sim}{v}}{|\underset{\sim}{u}|^2}\underset{\sim}{u}\right| = \left|(2\underset{\sim}{i} + \underset{\sim}{j}) - \frac{2}{5}(3\underset{\sim}{i} + 4\underset{\sim}{j})\right| = \left|\frac{5(2\underset{\sim}{i} + \underset{\sim}{j}) - 2(3\underset{\sim}{i} + 4\underset{\sim}{j})}{5}\right| = \left|\frac{10\underset{\sim}{i} + 5\underset{\sim}{j} - 6\underset{\sim}{i} - 8\underset{\sim}{j}}{5}\right| = \left|\frac{4\underset{\sim}{i} - 3\underset{\sim}{j}}{5}\right|$$

$$= \sqrt{\left(\frac{4}{5}\right)^2 + \left(\frac{-3}{5}\right)^2} = \sqrt{\frac{16}{25} + \frac{9}{25}} = \sqrt{\frac{25}{25}} = 1$$

$$\therefore A = \frac{1}{2}|\text{proj}_{\underset{\sim}{u}}\underset{\sim}{v}||\underset{\sim}{v} - \text{proj}_{\underset{\sim}{u}}\underset{\sim}{v}| = \frac{1}{2}(2)(1) = 1$$

Area of 2nd triangle: $A = \frac{1}{2}|\text{proj}_{\underset{\sim}{v}}\underset{\sim}{u}||\underset{\sim}{u} - \text{proj}_{\underset{\sim}{v}}\underset{\sim}{u}|$

$$|\text{proj}_{\underset{\sim}{v}}\underset{\sim}{u}| = \frac{\underset{\sim}{v}\cdot\underset{\sim}{u}}{|\underset{\sim}{v}|} = \frac{10}{\sqrt{5}} = \frac{10\sqrt{5}}{5} = 2\sqrt{5}$$

$\underset{\sim}{v}\cdot\underset{\sim}{u} = 10$ as above

$$|\underset{\sim}{v}| = \sqrt{2^2 + 1^2} = \sqrt{5}$$

$$\left|\underset{\sim}{u} - \text{proj}_{\underset{\sim}{v}}\underset{\sim}{u}\right| = \left|\underset{\sim}{u} - \frac{\underset{\sim}{v}\cdot\underset{\sim}{u}}{\left|\underset{\sim}{v}\right|^2}\underset{\sim}{v}\right|$$

$$= \left|(3\underset{\sim}{i} + 4\underset{\sim}{j}) - \frac{10}{(\sqrt{5})^2}(2\underset{\sim}{i} + \underset{\sim}{j})\right|$$

$$= \left|(3\underset{\sim}{i} + 4\underset{\sim}{j}) - 2(2\underset{\sim}{i} + \underset{\sim}{j})\right|$$

$$= \left|3\underset{\sim}{i} + 4\underset{\sim}{j} - 4\underset{\sim}{i} - 2\underset{\sim}{j}\right|$$

$$= \left|-\underset{\sim}{i} + 2\underset{\sim}{j}\right|$$

$$= \sqrt{(-1)^2 + 2^2}$$

$$= \sqrt{5}$$

$$\therefore A = \frac{1}{2}\left|\text{proj}_{\underset{\sim}{v}}\underset{\sim}{u}\right|\left|\underset{\sim}{u} - \text{proj}_{\underset{\sim}{v}}\underset{\sim}{u}\right|$$

$$= \frac{1}{2}(2\sqrt{5})(\sqrt{5})$$

$$= 5$$

As the projections were different, this makes their lengths different, hence the triangles and their areas are different.

Question 16

a $\underset{\sim}{F} = \begin{pmatrix} 3 \\ 6 \end{pmatrix}, \underset{\sim}{m} = \begin{pmatrix} 4 \\ 3 \end{pmatrix}$ ($\underset{\sim}{l}$ is parallel to $\underset{\sim}{m}$)

$\text{proj}_{\underset{\sim}{u}}\underset{\sim}{v} = \dfrac{\underset{\sim}{u}\cdot\underset{\sim}{v}}{\left|\underset{\sim}{u}\right|^2}\underset{\sim}{u}$ (projection of $\underset{\sim}{v}$ onto $\underset{\sim}{u}$)

$\text{proj}_{\underset{\sim}{m}}\underset{\sim}{F} = \dfrac{\underset{\sim}{m}\cdot\underset{\sim}{F}}{\left|\underset{\sim}{m}\right|^2}\underset{\sim}{m}$ (projection of $\underset{\sim}{F}$ onto $\underset{\sim}{m}$)

Step 1: $\underset{\sim}{m}\cdot\underset{\sim}{F} = (4)(3) + (6)(3)$
$= 30$

Step 2: $\left|\underset{\sim}{m}\right| = \sqrt{4^2 + 3^2}$
$= 5$

Step 3: $\text{proj}_{\underset{\sim}{m}}\underset{\sim}{F} = \dfrac{\underset{\sim}{m}\cdot\underset{\sim}{F}}{\left|\underset{\sim}{m}\right|^2}\underset{\sim}{m}$

$$= \frac{30}{5^2}(4\underset{\sim}{i} + 3\underset{\sim}{j})$$

$$\approx 4.8\underset{\sim}{i} + 3.6\underset{\sim}{j}$$

b To find the component form of $\underset{\sim}{F} \perp \underset{\sim}{l}$:

$\underset{\sim}{w} = \underset{\sim}{v} - \text{proj}_{\underset{\sim}{u}}\underset{\sim}{v}$ (orthogonal projection)

$$= \underset{\sim}{F} - \text{proj}_{\underset{\sim}{m}}\underset{\sim}{F}$$

$$= \begin{pmatrix} 3 \\ 6 \end{pmatrix} - \begin{pmatrix} 4.8 \\ 3.6 \end{pmatrix}$$

$$= \begin{pmatrix} -1.8 \\ 2.4 \end{pmatrix}$$

$$= -1.8\underset{\sim}{i} + 2.4\underset{\sim}{j}$$

Question 17

a Initial conditions: When $t = 0$:

$\dot{x} = 15\cos 55°$, $\dot{y} = 15\sin 55°$

$x = 0$ $y = 0$

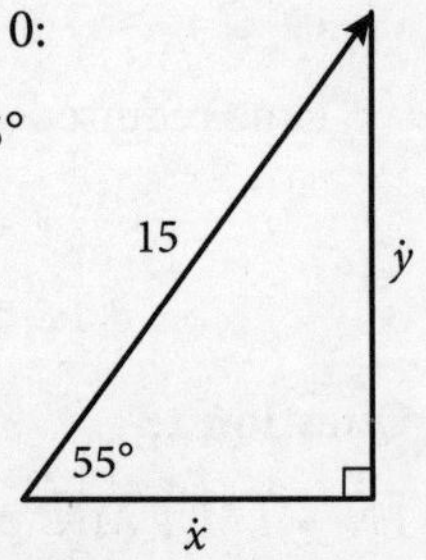

Determine equations of motion.

$\ddot{x} = 0$	$\ddot{y} = -10$
$\dot{x} = c_1$	$\dot{y} = -10t + c_3$
$t = 0, \dot{x} = 15\cos 55°$	$t = 0, \dot{y} = 15\sin 55°$
$c_1 = 15\cos 55°$	$c_3 = 15\sin 55°$
$\therefore \dot{x} = 15\cos 55°$	$\therefore \dot{y} = 15\sin 55° - 10t$
$x = 15t\cos 55° + c_2$	$y = 15t\sin 55° - 5t^2 + c_4$
$t = 0, x = 0$	$t = 0, y = 0$
$c_2 = 0$	$c_4 = 0$

$\therefore x = 15t\cos 55°$ [1]

$\therefore y = 15t\sin 55° - 5t^2$ [2]

When $x = 10$,

$$10 = 15t\cos 55°$$

$$t = \frac{10}{15\cos 55°}$$

Substitute into [2]:

$$y = 15t\sin 55° - 5t^2$$

$$= 15\sin 55° \times \frac{10}{15\cos 55°} - 5\left(\frac{10}{15\cos 55°}\right)^2$$

$$= 7.5268\ldots$$

$$= 7.5\text{ m}$$

b New equations of motion:

$\dot{x} = 5, \dot{y} = 0$:

Taking the origin to be at the point of rebound,

$x = 5t, y = -5t^2$

When $y = -7.5$:

$$-5t^2 = -7.5$$

$$t^2 = 1.5$$

$$t = 1.22\ldots \text{ seconds} \quad (t > 0)$$

Then $x = 5 \times 1.22\ldots$
$= 6.12\ldots$ m.

Distance from the thrower $= 10 - 6.12\ldots$
≈ 3.9 m.

Question 18

Let $\angle CAB = \theta$

$$|\overrightarrow{AB}| = \sqrt{3^2 + 1^2} = \sqrt{10}$$

$$\text{Area } \Delta ABC = \frac{ab}{2}\sin\theta$$

$$5 = \frac{|\overrightarrow{AB}|.|\overrightarrow{AC}|}{2}\sin\theta$$

$$5 = \frac{\sqrt{10}\,2\sqrt{5}}{2}\sin\theta$$

$$10 = 2\sqrt{50}\sin\theta$$

$$10 = 10\sqrt{2}\sin\theta$$

$$1 = \sqrt{2}\sin\theta$$

$$\frac{1}{\sqrt{2}} = \sin\theta$$

$$\theta = \sin^{-1}\left(\frac{1}{\sqrt{2}}\right)$$

$$\theta = \pm\frac{\pi}{4} \text{ or } \pm\frac{3\pi}{4}$$

$$\therefore \cos\theta = \pm\frac{1}{\sqrt{2}}$$

$$\underset{\sim}{a} \cdot \underset{\sim}{b} = |\underset{\sim}{a}||\underset{\sim}{b}|\cos\theta$$

$$\overrightarrow{AB} \cdot \overrightarrow{AC} = |\overrightarrow{AB}||\overrightarrow{AC}|\cos\theta$$

$$= \sqrt{10}\,2\sqrt{5} \times \pm\frac{1}{\sqrt{2}}$$

$$= \pm 10\sqrt{2} \times \frac{1}{\sqrt{2}}$$

$$= \pm 10$$

$$\overrightarrow{AB} \cdot \overrightarrow{AC} = \begin{pmatrix} 3 \\ 1 \end{pmatrix} \cdot \begin{pmatrix} x \\ y \end{pmatrix} \quad \rightarrow \quad \text{Let } \overrightarrow{AC} = \begin{pmatrix} x \\ y \end{pmatrix}$$

$$= (3)(x) + (1)(y)$$

$$= 3x + y$$

$$\therefore 3x + y = \pm 10$$

$$y = \pm 10 - 3x \qquad [1]$$

$$|\overrightarrow{AC}| = \sqrt{x^2 + y^2} = 2\sqrt{5}$$

$$\therefore x^2 + y^2 = \left(2\sqrt{5}\right)^2 = 20 \qquad [2]$$

Substitute [1] into [2]:

$$x^2 + (10 - 3x)^2 = 20$$
$$x^2 + (100 - 60x + 9x^2) = 20$$
$$10x^2 - 60x + 100 = 20$$
$$10x^2 - 60x + 80 = 0$$
$$x^2 - 6x + 8 = 0$$
$$(x - 2)(x - 4) = 0$$
$$x = 2 \text{ or } 4$$

OR

$$x^2 + (-10 - 3x)^2 = 20$$
$$x^2 + (100 + 60x + 9x^2) = 20$$
$$10x^2 + 60x + 100 = 20$$
$$10x^2 + 60x + 80 = 0$$
$$x^2 + 6x + 8 = 0$$
$$(x + 2)(x + 4) = 0$$
$$x = -2 \text{ or } -4$$

Substitute x into y: $y = 10 - 3(2)$, $y = 4$ OR $y = 10 - 3(4)$, $y = -2$ OR $y = -10 - 3(-2)$, $y = -4$ OR $y = 10 - 3(-4)$, $y = 2$

So possible $\overrightarrow{AC}$ vectors: $\begin{pmatrix} 2 \\ 4 \end{pmatrix}, \begin{pmatrix} 4 \\ -2 \end{pmatrix}, \begin{pmatrix} -2 \\ -4 \end{pmatrix}$ and $\begin{pmatrix} -4 \\ 2 \end{pmatrix}$.

Question 19

a $y = 50t\sin 88° - 5t^2$

$$= 50 \times 8 \times \sin 88° - 5(8)^2$$
$$= 79.75$$
$$\approx 80\,\text{m}$$

b Find speed of first skyrocket at $t = 8$.

$$\dot{x} = 50\cos 88°$$

$$\dot{y} = 50\sin 88° - 10t$$
$$= 50\sin 88° - 80$$

$$v = \sqrt{\dot{x}^2 + \dot{y}^2}$$
$$= \sqrt{(50\cos 88°)^2 + (50\sin 88° - 80)^2}$$
$$\approx 30 \text{ m s}^{-1}$$

Initial conditions: When $t = 0$:

$y = 80,\ \dot{y} = 15$

Determine vertical component of motion for second skyrocket.

$$\ddot{y} = -10$$
$$\dot{y} = -10t + c_1$$
$$t = 0, \dot{y} = 15$$
$$c_1 = 15$$

So $\dot{y} = 15 - 10t$

$$y = 15t - 5t^2 + c_2$$
$$t = 0, y = 80$$
$$c_2 = 80$$

So $y = 80 + 15t - 5t^2$.

The second skyrocket explodes when $\dot{y} = 0$.

$$0 = 15 - 10t$$
$$10t = 15$$
$$t = 1.5 \text{ s}$$

So $y = 80 + 15 \times 1.5 - 5(1.5)^2$

$$= 91.25$$
$$\approx 91 \text{ m}$$

Question 20

a $x = vt\cos\theta,\ y = vt\sin\theta - \frac{1}{2}gt^2$

When $y = 0$:

$$0 = \frac{t}{2}(2v\sin\theta - gt)$$
$$t = 0, \quad gt = 2v\sin\theta$$
$$t = 0, \quad t = \frac{2v\sin\theta}{g}$$

Water returns to the ground when $t = \frac{2v\sin\theta}{g}$.

Then $x = v\cos\theta \times \frac{2v\sin\theta}{g}$

$$= \frac{2v^2\sin\theta\cos\theta}{g}$$
$$= \frac{v^2\sin 2\theta}{g}$$

b From the information, when $\theta = 15°$, $x = 40$.

$$x = \frac{v^2\sin 2\theta}{g}$$
$$40 = \frac{v^2\sin 30°}{g}$$
$$= \frac{v^2}{2g}$$

So $v^2 = 80g$.

c $x = vt\cos\theta$, $\quad y = vt\sin\theta - \frac{1}{2}gt^2$ [2]

$t = \frac{x}{v\cos\theta}$ [1]

Substitute [1] into [2]:

$$y = v\sin\theta\left(\frac{x}{v\cos\theta}\right) - \frac{g}{2}\left(\frac{x}{v\cos\theta}\right)^2$$
$$= \frac{xv\sin\theta}{v\cos\theta} - \frac{gx^2}{2v^2\cos^2\theta} \qquad \left(\frac{1}{\cos^2\theta}\right) = \sec^2\theta$$
$$= x\tan\theta - \frac{gx^2\sec^2\theta}{2 \times 80g} \qquad (v^2 = 80g)$$
$$= x\tan\theta - \frac{x^2\sec^2\theta}{160}$$

d When $x = 40$, $y = 20$:

$$40\tan\theta - \frac{40^2\sec^2\theta}{160} = 20$$
$$40\tan\theta - 10\sec^2\theta = 20$$
$$4\tan\theta - (\tan^2\theta + 1) = 2$$
$$4\tan\theta - \tan^2\theta - 1 = 2$$
$$\tan^2\theta - 4\tan\theta + 3 = 0$$

e Solve equation above.

$$\tan^2\theta - 4\tan\theta + 3 = 0$$
$$(\tan\theta - 1)(\tan\theta - 3) = 0$$
$$\tan\theta = 1 \quad \text{OR} \quad \tan\theta = 3$$
$$\theta = 45° \qquad \theta = \tan^{-1}(3)$$

Water hits the top of the wall when $\theta = 45°, 71°34'$.

When the water hits the bottom of the wall, $x = 40$.

$$x = \frac{v^2\sin 2\theta}{g}$$
$$= 80\sin 2\theta \qquad v^2 = 80g$$
$$40 = 80\sin 2\theta$$
$$\sin 2\theta = \frac{1}{2}$$
$$2\theta = 30°, 150°$$
$$\therefore \quad \theta = 15°, 75°$$

The water hits the front of the wall when $15° < \theta < 45°$ and $72° < \theta < 75°$ (nearest degree).

HSC exam topic grid (2011–2020)

This table shows the coverage of this topic in past HSC exams by question number. The past exams can be downloaded from the NESA website (www.educationstandards.nsw.edu.au) by selecting 'Year 11 – Year 12', 'HSC exam papers'. NESA marking feedback and guidelines can also be found there.

The new Mathematics Extension 1 course was first examined in 2020. For exams before 2020, select 'Year 11 – Year 12', 'Resources archive', 'HSC exam papers archive'.

Vectors were introduced to the Mathematics Extension 1 course in 2020.

	Operations with vectors	Applying vectors	Projectile motion
2011			6(b)
2012			14(b)
2013			13(c)
2014			14(a)
2015			14(a)
2016			13(b)
2017			13(c)
2018			13(c)
2019			13(d)
2020 new course	**6**	**4**, **9**, 11(b)	

Questions in **bold** can be found in this chapter.

WORKED SOLUTIONS

CHAPTER 3
TRIGONOMETRIC EQUATIONS

ME-T3 Trigonometric equations 55

TRIGONOMETRIC EQUATIONS

Using trigonometric identities

- Compound angle identities
- Double angle identities
- Products to sums identities
- t-formulas
- Solving trigonometric equations

Auxiliary angle method

- $a\cos x + b\sin x = c$

Proving trigonometric identities

Glossary

A+ DIGITAL FLASHCARDS

Revise this topic's key terms and concepts by scanning the QR code or typing the URL into your browser.

https://get.ga/a-hsc-maths-ext-1

auxiliary angle method

A way of solving a trigonometric equation or expression written as the sum or difference of sin and cos by combining them into one sin or cos term, using the compound angle identities.

For example:

$$a\sin x + b\cos x = R\sin(x + \alpha)$$

$$R = \sqrt{a^2 + b^2} \text{ and } \tan\alpha = \frac{b}{a}$$

α is the auxiliary angle.

compound angle identities

Formulas for the trigonometric ratios of sums or differences of angles/values; for example:

$$\sin(A + B) = \sin A\cos B + \cos A\sin B$$

$$\cos(A - B) = \cos A\cos B + \sin A\sin B$$

domain

The set of possible values of x, the independent variable of a function or equation.

double angle identities

Formulas for the trigonometric ratios of doubles of angles/values, based on the compound angle identities above; for example:

$$\cos 2\theta = \cos^2\theta - \sin^2\theta$$

$$\tan 2\theta = \frac{2\tan\theta}{1 - \tan^2\theta}$$

identity

An equation that is always true no matter what values are substituted in. For example, $\cos^2 x + \sin^2 x = 1$ is a trigonometric identity.

products to sums identities

Formulas for converting products of sin and cos into sums of sin and cos:

$$\cos A\cos B = \frac{1}{2}[\cos(A - B) + \cos(A + B)]$$

$$\sin A\sin B = \frac{1}{2}[\cos(A - B) - \cos(A + B)]$$

$$\sin A\cos B = \frac{1}{2}[\sin(A + B) + \sin(A - B)]$$

$$\cos A\sin B = \frac{1}{2}[\sin(A + B) - \sin(A - B)]$$

t-formulas

Formulas for $\sin A$, $\cos A$ and $\tan A$ written in terms of $t = \tan\frac{A}{2}$:

$$\sin A = \frac{2t}{1 + t^2}$$

$$\cos A = \frac{1 - t^2}{1 + t^2}$$

$$\tan A = \frac{2t}{1 - t^2}$$

Topic summary

Trigonometric equations (ME-T3)

Compound angle identities

These identities are needed for differentiating, integrating, sketching and solving trigonometric functions.

$$\sin(A+B) = \sin A\cos B + \cos A\sin B$$

$$\sin(A-B) = \sin A\cos B - \cos A\sin B$$

$$\cos(A+B) = \cos A\cos B - \sin A\sin B$$

$$\cos(A-B) = \cos A\cos B + \sin A\sin B$$

$$\tan(A+B) = \frac{\tan A + \tan B}{1 - \tan A\tan B}$$

$$\tan(A-B) = \frac{\tan A - \tan B}{1 + \tan A\tan B}$$

Hint
Most of these identities appear on the HSC exam reference sheet and in the back of this book.

Double angle identities

These identities can be derived from the compound angle identities above by setting $A = B = \theta$.

$$\sin 2\theta = 2\sin\theta\cos\theta$$

$$\cos 2\theta = \cos^2\theta - \sin^2\theta$$

$$\tan 2\theta = \frac{2\tan\theta}{1 - \tan^2\theta}$$

From the $\cos 2\theta$ formula, formulas for $\cos^2\theta$ and $\sin^2\theta$ can be found:

$$\begin{aligned}\cos 2\theta &= \cos^2\theta - \sin^2\theta \\ &= 1 - \sin^2\theta - \sin^2\theta \quad (\cos^2\theta + \sin^2\theta = 1)\end{aligned}$$

$$\cos 2\theta = 1 - 2\sin^2\theta$$

$$2\sin^2\theta = 1 - \cos 2\theta$$

$$\sin^2\theta = \frac{1}{2}(1 - \cos 2\theta)$$

$$\begin{aligned}\cos 2\theta &= \cos^2\theta - \sin^2\theta \\ &= \cos^2\theta - (1 - \cos^2\theta) \quad (\cos^2\theta + \sin^2\theta = 1) \\ &= \cos^2\theta - 1 + \cos^2\theta\end{aligned}$$

$$\cos 2\theta = 2\cos^2\theta - 1$$

$$2\cos^2\theta = 1 + \cos 2\theta$$

$$\cos^2\theta = \frac{1}{2}(1 + \cos 2\theta)$$

The HSC exam reference sheet lists the more general formulas:

$$\sin^2 nx = \frac{1}{2}(1 - \cos 2nx)$$

$$\cos^2 nx = \frac{1}{2}(1 + \cos 2nx)$$

Products to sums identities

$$\cos A \cos B = \frac{1}{2}[\cos(A - B) + \cos(A + B)]$$

$$\sin A \sin B = \frac{1}{2}[\cos(A - B) - \cos(A + B)]$$

$$\sin A \cos B = \frac{1}{2}[\sin(A + B) + \sin(A - B)]$$

$$\cos A \sin B = \frac{1}{2}[\sin(A + B) - \sin(A - B)]$$

t-formulas

These formulas are used in solving trigonometric equations and proving identities.

If $t = \tan\frac{A}{2}$, then:

$$\sin A = \frac{2t}{1 + t^2}$$

$$\cos A = \frac{1 - t^2}{1 + t^2}$$

$$\tan A = \frac{2t}{1 - t^2}$$

$1 + t^2$

$2t$

t

$1 - t^2$

Important note: If $t = 180°$ is a solution, it will not be found by this method because $\tan\left(\frac{t}{2}\right) = \tan 90°$ is not defined. You will need to test separately whether $t = 180°$ is a solution to the given equation.

Auxiliary angle formulas

- These formulas are based on the compound angle identities and combine 2 trigonometric ratios into 1 (sine or cosine).
- To be used when specifically asked in this form; in this case, the *t*-formula is not to be used.

  Given $R = \sqrt{a^2 + b^2}$ and $\tan\alpha = \frac{b}{a}$:

$$a\sin x + b\cos x = R\sin(x + \alpha)$$

$$a\sin x - b\cos x = R\sin(x - \alpha)$$

$$a\cos x + b\sin x = R\cos(x - \alpha)$$

$$a\cos x - b\sin x = R\cos(x + \alpha)$$

Hint

These formulas are not on the HSC exam reference sheet, but they are often given in the HSC exam question.

These formulas can then be applied as a shifted wave function (**transformation**).

- For R, the amplitude and dilation factor, use $R = \sqrt{a^2 + b^2}$.
- For α, the auxiliary angle (phase shift), use $\tan\alpha = \frac{b}{a}$ and then find the angle in Quadrant 1 (if tan is positive) or Quadrant 2 (if tan is negative).

General solution of a trigonometric equation

This is required when there is no restriction on the **domain**.

Step 1: Find the solutions in the domain $0° \leq x \leq 360°$ or $0 \leq x \leq 2\pi$.

Step 2: Add multiples of $360°$ or 2π for sine and cosine and $180°$ or π for tangent, noting where m is an arbitrary integer. This is demonstrated in the solutions for Question 3 on page 68 and Question 7 on page 69.

Practice sets tracking grid

Maths is all about repetition, meaning do, do and do again! Each question in the following practice sets, especially the struggle questions (different for everybody!), should be completed at least 3 times correctly. Below is a tracking grid to record your question attempts: ✓ if you answered correctly, ✘ if you didn't.

PRACTICE SET 1: Multiple-choice questions

Question	1st attempt	2nd attempt	3rd attempt	4th attempt	5th attempt
1					
2					
3					
4					
5					
6					
7					
8					
9					
10					
11					
12					
13					
14					
15					
16					
17					
18					
19					
20					

PRACTICE SET 2: Short-answer questions

Question	1st attempt	2nd attempt	3rd attempt	4th attempt	5th attempt
1					
2					
3					
4					
5					
6					
7					
8					
9					
10					
11					
12					
13					
14					
15					
16					
17					
18					
19					
20					

Practice set 1

Multiple-choice questions

Solutions start on page 63.

Question 1

Which expression is equal to $\cos(A - B) - \cos(A + B)$?

A $-2\sin A\sin B$ **B** $-2\cos A\cos B$ **C** $2\sin A\sin B$ **D** $2\cos A\cos B$

Question 2

Which expression is equal to $\sec\theta$ if $t = \tan\frac{1}{2}\theta$?

A $\frac{2t}{1-t^2}$ **B** $\frac{1-t^2}{1+t^2}$ **C** $\frac{2t}{1+t^2}$ **D** $\frac{1+t^2}{1-t^2}$

Question 3

Which expression is equal to $2\sin 9\theta\cos 5\theta$?

A $\frac{1}{2}[\sin 14\theta + \sin 4\theta]$ **B** $\sin 9\theta - \sin 5\theta$ **C** $\frac{1}{2}[\sin 9\theta - \sin 5\theta]$ **D** $\sin 14\theta + \sin 4\theta$

Question 4

Which expression is equal to $2\sin x - \cos x$?

A $\sqrt{3}\sin(x - 30°)$ **B** $\sqrt{3}\sin(x - 26°34')$ **C** $\sqrt{5}\sin(x - 30°)$ **D** $\sqrt{5}\sin(x - 26°34')$

Question 5

Which expression is equal to $1 - \sin 2\theta$ if $t = \tan\theta$?

A $\frac{(t+1)^2}{1+t^2}$ **B** $\frac{(t-1)^2}{1+t^2}$ **C** $\frac{1-t^2}{1+t^2}$ **D** $\frac{1+t^2}{1-t^2}$

Question 6

Which expression is equal to $2\cos\left(\frac{7B}{2}\right)\sin\left(\frac{3B}{2}\right)$?

A $\frac{1}{2}[\sin 5B - \sin 2B]$ **B** $\sin 5B - \sin 2B$ **C** $\frac{1}{2}\left[\sin\frac{7B}{2} - \sin\frac{3B}{2}\right]$ **D** $\sin\frac{7B}{2} - \sin\frac{3B}{2}$

Question 7 ©NESA 2014 HSC EXAM, QUESTION 2

Which expression is equal to $\cos x - \sin x$?

A $\sqrt{2}\cos\left(x + \frac{\pi}{4}\right)$ **B** $\sqrt{2}\cos\left(x - \frac{\pi}{4}\right)$ **C** $2\cos\left(x + \frac{\pi}{4}\right)$ **D** $2\cos\left(x - \frac{\pi}{4}\right)$

Question 8

What is the value of $\sin 2\alpha$ given that $\tan\alpha = \sqrt{2}, 0 \le \alpha \le \frac{\pi}{2}$?

A $\frac{2\sqrt{2}}{3}$ **B** $\frac{3\sqrt{2}}{3}$ **C** $-\frac{3\sqrt{2}}{2}$ **D** $\frac{\sqrt{2}}{3}$

Question 9 ©NESA 1987 HSC EXAM, QUESTION 3(i)

For $0 \le x < 2\pi$, find all the solutions of the equation $\sin 2x = \cos x$.

A $0, \frac{\pi}{6}, \frac{5\pi}{6}, 2\pi$ **B** $\frac{\pi}{6}, \frac{7\pi}{6}, \frac{11\pi}{6}, 2\pi$ **C** $\frac{\pi}{2}, \frac{7\pi}{6}, \frac{3\pi}{2}, \frac{11\pi}{6}$ **D** $\frac{\pi}{6}, \frac{\pi}{2}, \frac{5\pi}{6}, \frac{3\pi}{2}$

Question 10

What is the value of $\tan 2x$ given $\sin x = \frac{3}{5}, \frac{\pi}{2} \le x \le \pi$?

A $-\frac{24}{7}$ **B** $-\frac{12}{7}$ **C** $\frac{12}{7}$ **D** $\frac{24}{7}$

Question 11

What is $\sin(A + B)$, given that $\sin A = t$, $180° < A < 270°$, $\cos B = t$, $90° < B < 180°$?

A $1 - 2t^2$ **B** $2t^2 - 1$ **C** 1 **D** −1

Question 12

What is the solution to $\tan 2x + \tan x = 0$ for $0° < x < 180°$?

A $x = 15°$ or $165°$ **B** $x = 30°$ or $150°$ **C** $x = 60°$ or $120°$ **D** $x = 75°$ or $105°$

Question 13

Which expression is equal to $4\sin\theta + 3\sin\theta + 5$ if $t = \tan\frac{\theta}{2}$?

A $\frac{2(t+2)^2}{1+t^2}$ **B** $\frac{(t+4)^2}{1+t^2}$ **C** $\frac{2(t+2)^2}{1-t^2}$ **D** $\frac{(t+4)^2}{1-t^2}$

Question 14

What is $\frac{\sin 2\theta + \sin\theta}{\cos 2\theta + \cos\theta + 1}$ in its simplest form?

A $\tan\theta$ **B** $\sin\theta$ **C** $\cot\theta$ **D** $\sec\theta$

Question 15 ©NESA 1990 HSC EXAM, QUESTION 5(a) MODIFIED FOR MULTIPLE-CHOICE

Find all the angles θ within $0 \le x \le 2\pi$, for which $\sin\theta + \cos\theta = 1$.

A $\theta = 0, \frac{\pi}{2}, 2\pi$ **B** $\theta = 0, \frac{\pi}{4}, \pi$ **C** $\theta = 0, \frac{\pi}{2}$ **D** $\theta = 0, \frac{\pi}{4}$

Question 16

Which expression is equal to $\cos\theta - \cos(\theta + 4\varphi)$?

A $2\sin(\theta + 4\varphi)\sin(\theta - 4\varphi)$ **B** $2\sin(\theta + 4\varphi)\sin(\theta - 2\varphi)$

C $2\sin(\theta + 2\varphi)\sin(2\varphi)$ **D** $2\sin(\theta - 4\varphi)\sin(\theta + 2\varphi)$

Question 17

Given $y = 3\sin^2\theta$ and $x = \sqrt{2\cos 2\theta}$, which of the following equations is correct?

A $y = \frac{3}{4}(2 - x^2)$ **B** $y = \frac{3}{4}(1 - x^2)$ **C** $y = \frac{3}{2}(2 - x^2)$ **D** $y = \frac{3}{2}(1 - x^2)$

PRACTICE SET 1

9780170459242

Question 18

Which expression is equal to $\sin x - \cos x$?

A $2\cos\left(x - \frac{\pi}{4}\right)$

B $\sqrt{2}\cos\left(x - \frac{3\pi}{4}\right)$

C $\sqrt{2}\cos\left(x - \frac{\pi}{4}\right)$

D $\sqrt{2}\cos\left(x + \frac{3\pi}{4}\right)$

Question 19

Which expression is equal to $\sin 3\theta$?

A $3\sin\theta - \sin^3\theta$

B $3\sin\theta - 4\sin^3\theta$

C $2\sin\theta - \sin^3\theta$

D $2\sin\theta - 4\sin^3\theta$

Question 20

What is $\sqrt{\frac{4 + 4\cos 2x}{1 - \cos 2x}}$ in its simplest form?

A $2\tan^2 x$

B $2\tan x$

C $2\cot x$

D $2\cot^2 x$

Practice set 2

Short-answer questions

Solutions start on page 68.

Question 1 (2 marks)

Prove that $\dfrac{1}{\sin^2\theta} + \dfrac{1}{\cos^2\theta} = \sec^2\theta\,\text{cosec}^2\theta$. 2 marks

Question 2 (2 marks)

Express $2\cos 75°\cos 15°$ as a sum or difference, then find its exact value. 2 marks

Question 3 (2 marks)

Find the general solution for $\sin 2x = \cos x$. 2 marks

Question 4 (1 mark)

Express $\sin\dfrac{7\pi}{24}\cos\dfrac{\pi}{24}$ as a sum or difference, then find its exact value. 1 mark

Question 5 (3 marks)

Express $\sin x - \sqrt{3}\cos x$ in the form $A\sin(x - \alpha)$, with $A > 0$ and $0 < \alpha < \dfrac{\pi}{2}$. 3 marks

Question 6 (3 marks)

Prove that $\dfrac{1 + \cos 2x}{\sin 2x} = \cot x$ and hence, or otherwise, find the exact value of $\cot 15°$. 3 marks

Question 7 (2 marks)

Solve $2\sin^2\theta - 1 = 0$. 2 marks

Question 8 (3 marks)

Solve $3\sin 2x = 2\tan x$ for $0° \le x \le 360°$. 3 marks

Question 9 (2 marks)

Simplify $\sin\left(\dfrac{\pi}{4} + x\right)\sin\left(\dfrac{\pi}{4} - x\right)$. 2 marks

Question 10 (2 marks)

Prove that $\dfrac{\sin 5\theta}{\sin\theta} - \dfrac{\cos 5\theta}{\cos\theta} = 4\cos 2\theta$. 2 marks

Question 11 (4 marks) ©NESA 2020 HSC EXAM, QUESTION 11(d)

By expressing $\sqrt{3}\sin x + 3\cos x$ in the form $A\sin(x + \alpha)$, solve $\sqrt{3}\sin x + 3\cos x = \sqrt{3}$, for $0 \le x \le 2\pi$. 4 marks

Question 12 (3 marks) ©NESA 1988 HSC EXAM, QUESTION 3(c)

Given that $0 < x < \dfrac{\pi}{4}$, prove that $\tan\left(\dfrac{\pi}{4} + x\right) = \dfrac{\cos x + \sin x}{\cos x - \sin x}$. 3 marks

Question 13 (4 marks)

a Express $\sqrt{3}\cos 2t - \sin 2t$ in the form $A\cos(2t + \alpha)$, with $A > 0$ and $0 < \alpha < \frac{\pi}{2}$. 2 marks

b Find, in exact form, the solution to $\sqrt{3}\cos 2t - \sin 2t = 1$, $0 < t < 2\pi$. 2 marks

Question 14 (3 marks)

Use $t = \tan\frac{x}{2}$ to solve $2\sin x + \cos x + 1 = 0$ for $0° \le x \le 360°$ to the nearest minute. 3 marks

Question 15 (2 marks)

Express $\frac{1}{2}[\sin 10A + \sin 4A]$ as a product of two trigonometric terms. 2 marks

Question 16 (2 marks)

Write $\sin(3\alpha + 5\beta)\sin(3\alpha - 5\beta)$ as a sum or difference of two trigonometric terms. 2 marks

Question 17 (2 marks) ©NESA 2010 HSC EXAM, QUESTION 6(a)

a Show that $\cos(A - B) = \cos A\cos B(1 + \tan A\tan B)$ 1 mark

b Suppose that $0 < B < \frac{\pi}{2}$ and $B < A < \pi$. 1 mark

Deduce that if $\tan A\tan B = -1$, then $A - B = \frac{\pi}{2}$.

Question 18 (4 marks)

The triangle ABC is isosceles, with $AB = BC = a$ and BD is perpendicular to AC.

Let $\angle ABD = \angle CBD = \theta$.

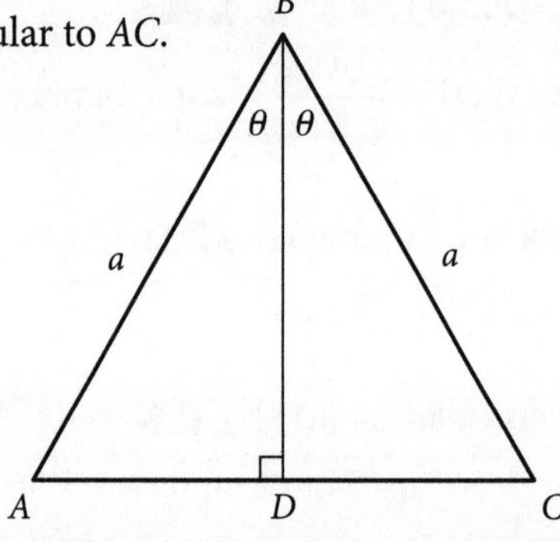

a Show that the area of ΔABD is $\frac{a^2\sin\theta\cos\theta}{2}$. 2 marks

b By considering the area of ΔABC, prove that $\sin 2\theta = 2\sin\theta\cos\theta$. 2 marks

Question 19 (4 marks) ©NESA 1995 HSC EXAM, QUESTION 5(a)

a Solve the equation $\sin 2x = 2\sin^2 x$ for $0 < x < \pi$. 2 marks

b Show that if $0 < x < \frac{\pi}{4}$, then $\sin 2x > 2\sin^2 x$. 2 marks

Question 20 (7 marks) ©NESA 2020 HSC EXAM, QUESTION 14(b)

a Show that $\sin^3\theta - \frac{3}{4}\sin\theta + \frac{\sin(3\theta)}{4} = 0$. 2 marks

b By letting $x = 4\sin\theta$ in the cubic equation $x^3 - 12x + 8 = 0$, show that $\sin(3\theta) = \frac{1}{2}$. 2 marks

c Prove that $\sin^2\frac{\pi}{18} + \sin^2\frac{5\pi}{18} + \sin^2\frac{25\pi}{18} = \frac{3}{2}$. 3 marks

Practice set 1

Hint
Remember that most of these formulas are on the HSC exam reference sheet.

Worked solutions

1 C

$$\frac{1}{2}[\cos(A - B) - \cos(A + B)] = \sin A \sin B$$
$$\cos(A - B) - \cos(A + B) = 2\sin A \sin B$$

2 D

$$\sec\theta = \frac{1}{\cos\theta} = \frac{1}{\frac{1 - t^2}{1 + t^2}} = \frac{1 + t^2}{1 - t^2}$$

3 D

$$\sin A \cos B = \frac{1}{2}[\sin(A + B) + \sin(A - B)]$$
$$2\sin A \cos B = \sin(A + B) + \sin(A - B)$$
$$2\sin 9\theta \cos 5\theta = \sin(9\theta + 5\theta) + \sin(9\theta - 5\theta) = \sin 14\theta + \sin 4\theta$$

4 D

Let $2\sin x - \cos x = R\sin(x - \alpha) = R(\sin x \cos\alpha - \cos x \sin\alpha) = R\sin x\cos\alpha - R\cos x \sin\alpha$

$\therefore R\cos\alpha = 2$ [1]

$\therefore R\sin\alpha = 1$ [2]

[2] ÷ [1]: $\tan\alpha = \frac{1}{2}$

$\alpha \approx 26°34'$

$[1]^2 + [2]^2$: $R^2\cos^2\alpha + R^2\sin^2\alpha = 2^2 + 1^2$

$R^2(\cos^2\alpha + \sin^2\alpha) = 5$

$R = \sqrt{5}$

So $2\sin x - \cos x = \sqrt{5}\sin(x - 26°34')$.

5 B

$$1 - \sin 2\theta = 1 - \frac{2t}{1 + t^2} = \frac{1 + t^2}{1 + t^2} - \frac{2t}{1 + t^2} = \frac{1 + t^2 - 2t}{1 + t^2} = \frac{(t - 1)^2}{1 + t^2}$$

6 B

$$\cos A \sin B = \frac{1}{2}[\sin(A + B) - \sin(A - B)]$$
$$2\cos A \sin B = \sin(A + B) - \sin(A - B)$$
$$2\cos\frac{7B}{2}\sin\frac{3B}{2} = \sin\left(\frac{7B}{2} + \frac{3B}{2}\right) - \sin\left(\frac{7B}{2} - \frac{3B}{2}\right) = \sin\frac{10B}{2} - \sin\frac{4B}{2} = \sin 5B - \sin 2B$$

7 A

$\cos x - \sin x = R\cos(x + \alpha) = R(\cos x\cos\alpha - \sin x \sin\alpha) = R\cos x\cos\alpha - R\sin x\sin\alpha$

$\therefore R\cos\alpha = 1$ [1]

$\therefore R\sin\alpha = 1$ [2]

[2] ÷ [1] $\tan\alpha = \frac{1}{1} = 1$

$\alpha = \frac{\pi}{4}$

$[1]^2 + [2]^2$: $R^2\cos^2\alpha + R^2\sin^2\alpha = 1^2 + 1^2$

$R^2(\cos^2\alpha + \sin^2\alpha) = 2$

$R = \sqrt{2}$

So $\cos x - \sin x = \sqrt{2}\cos\left(x + \frac{\pi}{4}\right)$.

8 A

$\tan\alpha = \sqrt{2}$

$\tan\alpha = \frac{\sqrt{2}}{1}$

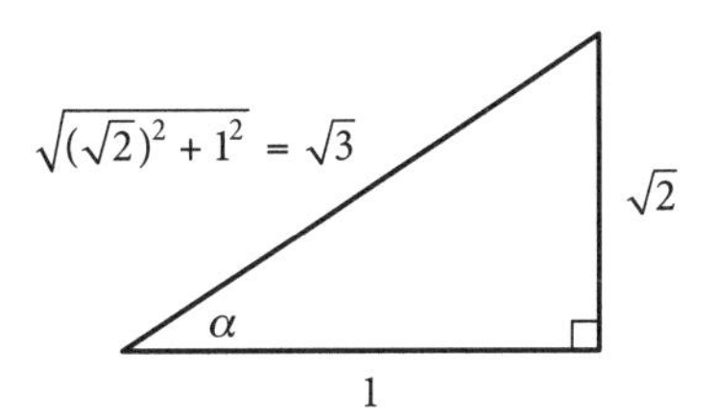

$$\sin 2\alpha = 2\sin\alpha\cos\alpha = 2\left(\frac{\sqrt{2}}{\sqrt{3}}\right)\left(\frac{1}{\sqrt{3}}\right) = \frac{2\sqrt{2}}{3}$$

WORKED SOLUTIONS

9 D

$$\sin 2x = \cos x$$
$$2\sin x\cos x = \cos x$$
$$2\sin x\cos x - \cos x = 0$$
$$\cos x(2\sin x - 1) = 0$$

$\cos x = 0$ OR $2\sin x - 1 = 0$

$x = \cos^{-1}(0)$ $\qquad 2\sin x = 1$

$x = \dfrac{\pi}{2}, \dfrac{3\pi}{2}$ $\qquad \sin x = \dfrac{1}{2}$

$x = \sin^{-1}\left(\dfrac{1}{2}\right)$

$x = \dfrac{\pi}{6}, \dfrac{5\pi}{6}$

So $x = \dfrac{\pi}{6}, \dfrac{\pi}{2}, \dfrac{5\pi}{6}, \dfrac{3\pi}{2}$.

10 A

$\sin x = \dfrac{3}{5}$

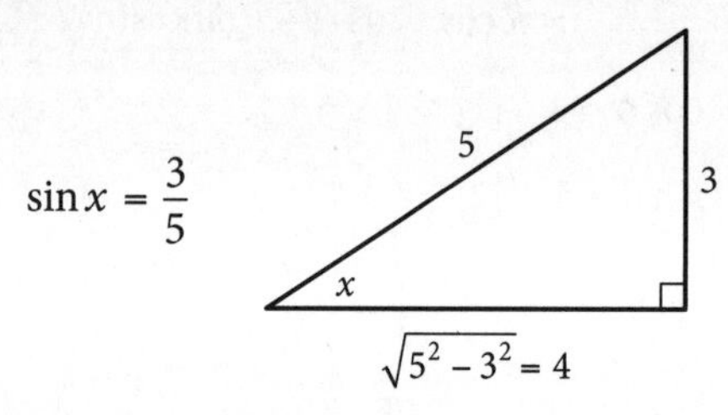

$$\tan 2x = \frac{2\tan x}{1-\tan^2 x}$$

$\tan x < 0$ in 2nd quadrant, so $\tan x = -\dfrac{3}{4}$.

$$= \frac{2\left(-\frac{3}{4}\right)}{1-\left(-\frac{3}{4}\right)^2}$$
$$= \frac{-\frac{3}{2}}{\frac{7}{16}}$$
$$= -\frac{24}{7}$$

11 B

$\sin A = t$: Quadrant 3 $\qquad \cos B = t$: Quadrant 2

$$\sin(A+B) = \sin A\cos B + \cos A\sin B = t \times t + \left(-\frac{\sqrt{1-t^2}}{1}\right) \times \left(\frac{\sqrt{1-t^2}}{1}\right)$$
$$= t^2 - (1-t^2)$$
$$= 2t^2 - 1$$

12 C

$$\tan 2x + \tan x = 0$$

$$\frac{2\tan x}{1 - \tan^2 x} + \tan x = 0$$

$$\tan x\left(\frac{2}{1 - \tan^2 x} + 1\right) = 0$$

$$\tan x = 0$$

$$x = 0 \text{ but not in domain}$$

OR

$$\frac{2}{1 - \tan^2 x} + 1 = 0$$

$$\frac{2}{1 - \tan^2 x} = -1$$

$$2 = -1 + \tan^2 x$$

$$3 = \tan^2 x$$

$$\pm\sqrt{3} = \tan x$$

$$x = 60°, 120°$$

13 A

$$4\sin\theta + 3\cos\theta + 5$$

$$= 4 \times \frac{2t}{1 + t^2} + 3 \times \frac{1 - t^2}{1 + t^2} + 5$$

$$= \frac{8t}{1 + t^2} + \frac{3 - 3t^2}{1 + t^2} + 5$$

$$= \frac{8t + 3 - 3t^2}{1 + t^2} + \frac{5(1 + t^2)}{1 + t^2}$$

$$= \frac{8t + 3 - 3t^2 + 5 + 5t^2}{1 + t^2}$$

$$= \frac{2t^2 + 8t + 8}{1 + t^2}$$

$$= \frac{2(t^2 + 4t + 4)}{1 + t^2}$$

$$= \frac{2(t + 2)^2}{1 + t^2}$$

14 A

$$\frac{\sin 2\theta + \sin\theta}{\cos 2\theta + \cos\theta + 1}$$

$$= \frac{2\sin\theta\cos\theta + \sin\theta}{\cos 2\theta + \cos\theta + 1}$$

$$= \frac{\sin\theta(2\cos\theta + 1)}{\cos 2\theta + \cos\theta + 1}$$

$$= \frac{\sin\theta(2\cos\theta + 1)}{2\cos^2\theta - 1 + \cos\theta + 1}$$

$$= \frac{\sin\theta(2\cos\theta + 1)}{2\cos^2\theta + \cos\theta}$$

$$= \frac{\sin\theta(2\cos\theta + 1)}{\cos\theta(2\cos\theta + 1)}$$

$$= \frac{\sin\theta}{\cos\theta}$$

$$= \tan\theta$$

WORKED SOLUTIONS

15 A

Let $t = \tan\frac{1}{2}\theta$ (Use t-formulas: $\sin A = \frac{2t}{1+t^2}$, $\cos A = \frac{1-t^2}{1+t^2}$)

$\sin\theta + \cos\theta = 1$

$$\frac{2t}{1+t^2} + \frac{1-t^2}{1+t^2} = 1 \quad \text{(provided } \theta \neq \pi)$$

$$\frac{2t+1-t^2}{1+t^2} = 1$$

$$2t + 1 - t^2 = 1 + t^2$$

$$0 = 2t^2 - 2t$$

$$0 = t^2 - t$$

$$0 = t(t-1)$$

$$t = 0 \text{ or } 1$$

Hint

The auxiliary angle method also works here, resulting in the equation $\sqrt{2}\sin\left(x + \frac{\pi}{4}\right) = 1$.

$$t = \tan\frac{1}{2}\theta \quad \left(0 \leq \frac{1}{2}\theta \leq \pi\right)$$

$\tan\frac{1}{2}\theta = 0$ OR $\tan\frac{1}{2}\theta = 1$

$\frac{1}{2}\theta = 0, \pi$ $\qquad$ $\frac{1}{2}\theta = \frac{\pi}{4}$

$\theta = 0, 2\pi$ $\qquad$ $\theta = \frac{\pi}{2}$

Note: With t-formulas, always check if $180° = \pi$ is a solution.

$\sin\pi + \cos\pi = 0 + (-1) \neq 1$

$\therefore\ \theta = \pi$ is not a solution.

So $\theta = 0, \frac{\pi}{2}, 2\pi$.

16 C

Notice that all possible solutions involve a product of sines, so we know to use this formula:

$$\sin A \sin B = \frac{1}{2}[\cos(A-B) - \cos(A+B)].$$

$2\sin A \sin B = \cos(A-B) - \cos(A+B)$

In $\cos\theta - \cos(\theta + 4\varphi)$:

$A - B = \theta$ [1]

$A + B = \theta + 4\varphi$ [2]

[1] + [2]:

$$2A = 2\theta + 4\varphi$$

$$A = \theta + 2\varphi$$

Substitute $A = \theta + 2\varphi$ into [1]:

$$(\theta + 2\varphi) - B = \theta$$

$$2\varphi - B = 0$$

$$B = 2\varphi$$

So $\cos\theta - \cos(\theta + 4\varphi) = 2\sin(\theta + 2\varphi)\cos(2\varphi)$.

17 A

$$x = \sqrt{2\cos 2\theta}$$
$$x^2 = 2\cos 2\theta$$
$$x^2 = 2(\cos^2\theta - \sin^2\theta)$$
$$x^2 = 2\left[(1 - \sin^2\theta) - \sin^2\theta\right]$$
$$x^2 = 2\left[1 - 2\sin^2\theta\right]$$
$$x^2 = 2 - 4\sin^2\theta$$
$$x^2 - 2 = -4\sin^2\theta$$
$$\frac{x^2 - 2}{-4} = \sin^2\theta$$
$$\frac{2 - x^2}{4} = \sin^2\theta$$

$$\begin{aligned} y &= 3\sin^2\theta \\ &= 3\left(\frac{2 - x^2}{4}\right) \\ &= \frac{3}{4}(2 - x^2) \end{aligned}$$

18 B

$$\begin{aligned} \sin x - \cos x &= R\cos(x - \alpha) \\ &= R(\cos x\cos\alpha + \sin x\sin\alpha) \\ -\cos x + \sin x &= R\cos x\cos\alpha + R\sin x\sin\alpha \end{aligned}$$

$\therefore R\cos\alpha = -1$ [1]

$\therefore R\sin\alpha = 1$ [2]

[2] ÷ [1]: $\tan\alpha = \frac{1}{-1} = -1$

According to [1] and [2], $\cos\alpha$ is negative and $\sin\alpha$ is positive: 2nd quadrant.

$$\alpha = \frac{3\pi}{4}$$

$[1]^2 + [2]^2$: $R^2\cos^2\alpha + R^2\sin^2\alpha = (-1)^2 + 1^2$

$$R^2(\cos^2\alpha + \sin^2\alpha) = 2$$
$$R = \sqrt{2}$$

So $\sin x - \cos x = \sqrt{2}\cos\left(x - \frac{3\pi}{4}\right)$.

19 B

$$\begin{aligned} \sin 3\theta &= \sin(2\theta + \theta) \\ &= \sin 2\theta\cos\theta + \cos 2\theta\sin\theta && (\sin(A + B) = \sin A\cos B + \cos A\sin B) \\ &= 2\sin\theta\cos\theta\cos\theta + \cos 2\theta\sin\theta && (\sin 2\theta = 2\sin\theta\cos\theta) \\ &= 2\sin\theta\cos^2\theta + (\cos^2\theta - \sin^2\theta)\sin\theta && (\cos 2\theta = \cos^2\theta - \sin^2\theta) \\ &= 2\sin\theta\cos^2\theta + \sin\theta\cos^2\theta - \sin^3\theta \\ &= 3\sin\theta\cos^2\theta - \sin^3\theta \\ &= 3\sin\theta(1 - \sin^2\theta) - \sin^3\theta && (\cos^2\theta = 1 - \sin^2\theta) \\ &= 3\sin\theta - 3\sin^3\theta - \sin^3\theta \\ &= 3\sin\theta - 4\sin^3\theta \end{aligned}$$

20 C

$$\begin{aligned} \sqrt{\frac{4 + 4\cos 2x}{1 - \cos 2x}} &= \frac{\sqrt{4}\left(\sqrt{1 + \cos 2x}\right)}{\sqrt{1 - \cos 2x}} \\ &= \frac{2\sqrt{1 + \cos 2x}}{\sqrt{1 - \cos 2x}} \\ &= 2\sqrt{\frac{1 + 2\cos^2 x - 1}{1 - \cos 2x}} && (\cos 2x = 2\cos^2 x - 1) \\ &= 2\sqrt{\frac{2\cos^2 x}{1 - (1 - 2\sin^2 x)}} && (\cos 2x = 1 - 2\sin^2 x) \\ &= 2\sqrt{\frac{2\cos^2 x}{2\sin^2 x}} \\ &= 2\sqrt{\frac{\cos^2 x}{\sin^2 x}} \\ &= 2\sqrt{\cot^2 x} \\ &= 2\cot x \end{aligned}$$

Practice set 2

Worked solutions

Question 1

$$\frac{1}{\sin^2\theta}+\frac{1}{\cos^2\theta}=\frac{1(\cos^2\theta)+1(\sin^2\theta)}{\sin^2\theta\cos^2\theta}$$
$$=\frac{\cos^2\theta+\sin^2\theta}{\sin^2\theta\cos^2\theta}$$
$$=\frac{1}{\sin^2\theta\cos^2\theta}$$
$$=\frac{1}{\sin^2\theta}\times\frac{1}{\cos^2\theta}$$
$$=\operatorname{cosec}^2\theta\times\sec^2\theta$$
$$=\sec^2\theta\operatorname{cosec}^2\theta$$

Question 2

$$\cos A\cos B=\frac{1}{2}[\cos(A+B)+\cos(A-B)]$$
$$2\cos A\cos B=\cos(A+B)+\cos(A-B)$$
$$2\cos 75^\circ\cos 15^\circ=\cos(75^\circ+15^\circ)+\cos(75^\circ-15^\circ)$$
$$=\cos 90^\circ+\cos 60^\circ$$
$$=0+\frac{1}{2}$$
$$=\frac{1}{2}$$

Question 3

$\sin 2x=\cos x$

$2\sin x\cos x=\cos x$

$2\sin x\cos x-\cos x=0$

$\cos x(2\sin x-1)=0$

$\cos x=0$ OR $2\sin x-1=0$

$x=\cos^{-1}(0)$

$x=\frac{\pi}{2},\frac{3\pi}{2},\ldots$

$x=\frac{\pi}{2}+\pi m$

Hint
We can write the answer this way if the solutions have a common difference of π.

$2\sin x=1$

$\sin x=\frac{1}{2}$ (Quadrants 1 and 2)

$x=\sin^{-1}\left(\frac{1}{2}\right)$

$x=\frac{\pi}{6},\pi-\frac{\pi}{6},\ldots$

$x=\frac{\pi}{6},\frac{5\pi}{6},\ldots$

$x=\frac{\pi}{6}+2\pi m$ or $\frac{5\pi}{6}+2\pi m$

So $x=\frac{\pi}{2}+\pi m$ or $\frac{\pi}{6}+2\pi m$ or $\frac{5\pi}{6}+2\pi m$, where m is an arbitrary integer.

Note: The solution must be in general form because no domain was given.

Question 4

$$\sin A \cos B = \frac{1}{2}[\sin(A+B) + \sin(A-B)]$$

$$\begin{aligned}\sin\frac{7\pi}{24}\cos\frac{\pi}{24} &= \frac{1}{2}\left[\sin\left(\frac{7\pi}{24}+\frac{\pi}{24}\right) + \sin\left(\frac{7\pi}{24}-\frac{\pi}{24}\right)\right]\\ &= \frac{1}{2}\left[\sin\frac{8\pi}{24} + \sin\frac{6\pi}{24}\right]\\ &= \frac{1}{2}\left[\sin\frac{\pi}{3} + \sin\frac{\pi}{4}\right]\\ &= \frac{1}{2}\left[\frac{\sqrt{3}}{2} + \frac{1}{\sqrt{2}}\right]\\ &= \frac{1}{2}\left[\frac{\sqrt{3}}{2} + \frac{\sqrt{2}}{2}\right]\\ &= \frac{1}{2}\left[\frac{\sqrt{3}+\sqrt{2}}{2}\right]\\ &= \frac{\sqrt{3}+\sqrt{2}}{4}\end{aligned}$$

Question 5

$$\begin{aligned}\sin x - \sqrt{3}\cos x &= R\sin(x-\alpha)\\ &= R(\sin x\cos\alpha - \cos x\sin\alpha)\\ &= R\sin x\cos\alpha - R\cos x\sin\alpha\end{aligned}$$

$\therefore R\cos\alpha = 1$ [1]

$\therefore R\sin\alpha = \sqrt{3}$ [2]

[2] ÷ [1]: $\tan\alpha = \frac{\sqrt{3}}{1} = \sqrt{3}$

$\alpha = \frac{\pi}{3}$

$[1]^2 + [2]^2$:

$$\begin{aligned}R^2\cos^2\alpha + R^2\sin^2\alpha &= 1^2 + \left(\sqrt{3}\right)^2\\ R^2(\cos^2\alpha + \sin^2\alpha) &= 4\\ R &= 2\end{aligned}$$

So $\sin x - \sqrt{3}\cos x = 2\sin\left(x - \frac{\pi}{3}\right)$.

Question 6

$$\begin{aligned}\frac{1+\cos 2x}{\sin 2x} &= \frac{2\cos^2 x}{2\sin x\cos x}\\ &= \frac{\cos x}{\sin x}\\ &= \cot x\end{aligned}$$

$$\begin{aligned}\text{So } \cot 15° &= \frac{1+\cos(2\times 15°)}{\sin(2\times 15°)}\\ &= \frac{1+\cos 30°}{\sin 30°}\\ &= \frac{1+\left(\frac{\sqrt{3}}{2}\right)}{\frac{1}{2}}\\ &= 2 + 2\left(\frac{\sqrt{3}}{2}\right)\\ &= 2+\sqrt{3}\end{aligned}$$

Question 7

$$\begin{aligned}2\sin^2\theta - 1 &= 0\\ 2\sin^2\theta &= 1\\ \sin^2\theta &= \frac{1}{2}\\ \sin\theta &= \pm\sqrt{\frac{1}{2}} \qquad \text{(All quadrants)}\\ \theta &= \frac{\pi}{4}, \frac{3\pi}{4}, \ldots\end{aligned}$$

The solution is $\theta = \frac{\pi}{4} + \frac{m}{2}$, where m is an arbitrary integer.

Hint

The solution must be in general form because no domain was given. We can write it this way because the values differ by a constant $\frac{\pi}{2}$.

Question 8

$$\begin{aligned}3\sin 2x &= 2\tan x\\ 3(2\sin x\cos x) &= 2\tan x\\ 6\sin x\cos x &= \frac{2\sin x}{\cos x}\\ 6\sin x\cos^2 x &= 2\sin x\\ 6\sin x\cos^2 x - 2\sin x &= 0\\ 2\sin x(3\cos^2 x - 1) &= 0\end{aligned}$$

$2\sin x = 0$

$x = 0°, 180°, 360°$

OR $3\cos^2 x - 1 = 0$

$$\begin{aligned}\cos^2 x &= \frac{1}{3}\\ \cos x &= \pm\sqrt{\frac{1}{3}} \quad \text{(All quadrants)}\\ x &= \cos^{-1}\left(\frac{1}{\sqrt{3}}\right)\\ x &= 54.7°, 180° - 54.7°, 180° + 54.7°, 360° - 54.7°\\ x &= 54.7°, 125.3°, 234.7°, 305.3°, \ldots\end{aligned}$$

So $x = 0°, 54.7°, 125.3°, 180°, 234.7°, 305.3°, 360°$.

Question 9

$\sin A\sin B = \frac{1}{2}\left[\cos(A - B) - \cos(A + B)\right]$

$$\begin{aligned}\sin\left(\frac{\pi}{4} + x\right)\sin\left(\frac{\pi}{4} - x\right) &= \frac{1}{2}\left[\cos\left(\left(\frac{\pi}{4} + x\right) - \left(\frac{\pi}{4} - x\right)\right) - \cos\left(\left(\frac{\pi}{4} + x\right) + \left(\frac{\pi}{4} - x\right)\right)\right]\\ &= \frac{1}{2}\left[\cos(2x) - \cos\left(\frac{2\pi}{4}\right)\right]\\ &= \frac{1}{2}\left[\cos 2x - \cos\frac{\pi}{2}\right]\\ &= \frac{1}{2}\left[\cos 2x - 0\right]\\ &= \frac{1}{2}\cos 2x\end{aligned}$$

Question 10

$$\begin{aligned}\frac{\sin 5\theta}{\sin\theta} - \frac{\cos 5\theta}{\cos\theta} &= \frac{\sin 5\theta \times \cos\theta}{\sin\theta \times \cos\theta} - \frac{\cos 5\theta \times \sin\theta}{\cos\theta \times \sin\theta}\\ &= \frac{\sin 5\theta\cos\theta - \cos 5\theta\sin\theta}{\sin\theta\cos\theta}\\ &= \frac{\sin 5\theta\cos\theta - \cos 5\theta\sin\theta}{\sin\theta\cos\theta}\\ &= \frac{\sin(5\theta - \theta)}{\sin\theta\cos\theta} \qquad \sin(A - B) = \sin A\cos B - \cos A\sin B\\ &= \frac{\sin 4\theta}{\sin\theta\cos\theta}\\ &= \frac{2\sin 2\theta\cos 2\theta}{\frac{1}{2}\sin 2\theta} \qquad \sin 2x = 2\sin x\cos x\\ &= \frac{4\sin 2\theta\cos 2\theta}{\sin 2\theta}\\ &= 4\cos 2\theta\end{aligned}$$

Question 11

$$\begin{aligned}\sqrt{3}\sin x + 3\cos x &= R\sin(x+\alpha)\\ &= R(\sin x\cos\alpha + \cos x\sin\alpha)\\ &= R\sin x\cos\alpha + R\cos x\sin\alpha\end{aligned}$$

$\therefore R\cos\alpha = \sqrt{3}$ [1]

$\therefore R\sin\alpha = 3$ [2]

[2] ÷ [1]: $\tan\alpha = \dfrac{3}{\sqrt{3}} = \sqrt{3}$

$\alpha = \dfrac{\pi}{3}$

$[1]^2 + [2]^2$: $R^2\cos^2\alpha + R^2\sin^2\alpha = \left(\sqrt{3}\right)^2 + 3^2$

$R^2(\cos^2\alpha + \sin^2\alpha) = 12$

$R = \sqrt{12} = 2\sqrt{3}$

So $\sqrt{3}\sin x + 3\cos x = 2\sqrt{3}\sin\left(x + \dfrac{\pi}{3}\right)$.

$$\begin{aligned}\text{So } 2\sqrt{3}\sin\left(x+\frac{\pi}{3}\right) &= \sqrt{3}\\ \sin\left(x+\frac{\pi}{3}\right) &= \frac{1}{2}\\ x + \frac{\pi}{3} &= \sin^{-1}\left(\frac{1}{2}\right)\\ x + \frac{\pi}{3} &= \frac{\pi}{6}, \pi - \frac{\pi}{6}, 2\pi + \frac{\pi}{6}\\ x + \frac{\pi}{3} &= \frac{\pi}{6}, \frac{5\pi}{6}, \frac{13\pi}{6}\\ x &= \frac{\pi}{6} - \frac{\pi}{3}, \frac{5\pi}{6} - \frac{\pi}{3}, \frac{13\pi}{6} - \frac{\pi}{3}\\ x &= -\frac{\pi}{6}, \frac{\pi}{2}, \frac{11\pi}{6}\end{aligned}$$

So $x = \dfrac{\pi}{2}, \dfrac{11\pi}{6}$ given $0 \le x \le 2\pi$.

Question 12

$$\tan(A+B) = \frac{\tan A + \tan B}{1 - \tan A\tan B}$$

$$\begin{aligned}\tan\left(\frac{\pi}{4} + x\right) &= \frac{\tan\frac{\pi}{4} + \tan x}{1 - \tan\frac{\pi}{4}\tan x}\\ &= \frac{1 + \tan x}{1 - \tan x}\\ &= \frac{1 + \frac{\sin x}{\cos x}}{1 - \frac{\sin x}{\cos x}}\\ &= \frac{1 + \frac{\sin x}{\cos x}}{1 - \frac{\sin x}{\cos x}} \times \frac{\cos x}{\cos x}\\ &= \frac{\cos x + \sin x}{\cos x - \sin x}\end{aligned}$$

Question 13

a $\sqrt{3}\cos 2t - \sin 2t = R\cos(2t + \alpha)$

$= R(\sin 2t\cos\alpha - \cos 2t\sin\alpha)$

$= R\sin 2t\cos\alpha - R\cos 2t\sin\alpha$

$\therefore R\cos\alpha = \sqrt{3}$ [1]

$\therefore R\sin\alpha = 1$ [2]

[2] ÷ [1]: $\tan\alpha = \dfrac{1}{\sqrt{3}}$

$\alpha = \dfrac{\pi}{6}$

$[1]^2 + [2]^2$: $R^2\cos^2\alpha + R^2\sin^2\alpha = (\sqrt{3})^2 + 1^2$

$R^2(\cos^2\alpha + \sin^2\alpha) = 4$

$R = 2$

So $\sqrt{3}\cos 2t - \sin 2t = 2\cos\left(2t + \dfrac{\pi}{6}\right)$.

b $2\cos\left(2t + \dfrac{\pi}{6}\right) = 1$

$\cos\left(2t + \dfrac{\pi}{6}\right) = \dfrac{1}{2}$ (Quadrants 1 and 4)

$2t + \dfrac{\pi}{6} = \cos^{-1}\left(\dfrac{1}{2}\right)$

$2t + \dfrac{\pi}{6} = \dfrac{\pi}{3}, 2\pi + \dfrac{\pi}{3}, 2\pi - \dfrac{\pi}{3}, 4\pi + \dfrac{\pi}{3}, 4\pi - \dfrac{\pi}{3}, \ldots$

$2t + \dfrac{\pi}{6} = \dfrac{\pi}{3}, \dfrac{7\pi}{3}, \dfrac{5\pi}{3}, \dfrac{13\pi}{3}, \dfrac{11\pi}{3}, \ldots$

$2t = \dfrac{\pi}{3} - \dfrac{\pi}{6}, \dfrac{7\pi}{3} - \dfrac{\pi}{6}, \dfrac{5\pi}{3} - \dfrac{\pi}{6}, \dfrac{13\pi}{3} - \dfrac{\pi}{6}, \dfrac{11\pi}{3} - \dfrac{\pi}{6}, \ldots$

$2t = \dfrac{\pi}{6}, \dfrac{13\pi}{6}, \dfrac{3\pi}{2}, \dfrac{25\pi}{6}, \dfrac{7\pi}{2}, \ldots$

$t = \dfrac{\pi}{12}, \dfrac{13\pi}{12}, \dfrac{3\pi}{4}, \dfrac{25\pi}{12}, \dfrac{7\pi}{4}, \ldots$

So $t = \dfrac{\pi}{12}, \dfrac{3\pi}{4}, \dfrac{13\pi}{12}, \dfrac{7\pi}{4}$. $(0 < t < 2\pi)$

Question 14

$$2\sin x + \cos x + 1 = 0$$

$$2 \times \frac{2t}{1+t^2} + \frac{1-t^2}{1+t^2} = -1 \qquad \text{(provided } x \neq 180°\text{)}$$

$$\frac{4t+1-t^2}{1+t^2} = -1$$

$$4t + 1 - t^2 = -1 - t^2$$

$$4t + 2 = 0$$

$$2(2t+1) = 0$$

$$2t + 1 = 0$$

$$2t = -1$$

$$t = -\frac{1}{2}$$

Hint
We could also use the auxiliary angle method here, resulting in the equation $\sqrt{5}\sin(x + 26°35') = -1$.

$$t = \tan\frac{1}{2}x \qquad \left(0° \le \frac{1}{2}x \le 180°\right)$$

$$\tan\frac{1}{2}x = -\frac{1}{2} \qquad \text{(Quadrants 2 and 4)}$$

$$\frac{1}{2}x = \tan^{-1}\left(-\frac{1}{2}\right)$$

$$= 180° - 26.565\,05\ldots°$$

$$= 153.4349\ldots°$$

$$x \approx 306°52'$$

Hint
Do not round the answer until the end!

Check if 180° is a solution.

$$2\sin 180° + \cos 180° + 1 = 0 + (-1) + 1 = 0$$

$\therefore$ 180° is a solution.

So $x = 180°, 306°52'$.

Question 15

$$\sin A\cos B = \frac{1}{2}\left[\sin(A+B) + \sin(A-B)\right]$$

$A + B = 10$ [1]

$A - B = 4$ [2]

[1] + [2]: $2A = 14$
$A = 7$

Substitute $A = 7$ into [1]: $7 + B = 10$
$B = 3$

So $\frac{1}{2}[\sin 10A + \sin 4A] = \sin 7A\cos 3A$.

Question 16

$$\sin A\sin B = \frac{1}{2}\left[\cos(A-B) - \cos(A+B)\right]$$

$$\sin(3\alpha + 5\beta)\sin(3\alpha - 5\beta) = \frac{1}{2}\left[\cos\left((3\alpha+5\beta) - (3\alpha-5\beta)\right) - \cos\left((3\alpha+5\beta) + (3\alpha-5\beta)\right)\right]$$

$$= \frac{1}{2}\left[\cos(3\alpha + 5\beta - 3\alpha + 5\beta) - \cos(3\alpha + 5\beta + 3\alpha - 5\beta)\right]$$

$$= \frac{1}{2}\left[\cos(10\beta) - \cos(6\alpha)\right]$$

Question 17

a $\cos(A - B)$

$= \cos A \cos B + \sin A \sin B$

$= \cos A \cos B + \sin A \sin B \times \dfrac{\cos A \cos B}{\cos A \cos B}$

$= \cos A \cos B\left(1 + \dfrac{\sin A \sin B}{\cos A \cos B}\right)$

$= \cos A \cos B(1 + \tan A \tan B)$

b If $\tan A \tan B = -1$, then:

$\cos(A - B)$

$= \cos A \cos B(1 + (-1))$

$= \cos A \cos B(0)$

$= 0$

$A > B > 0$ and $A < \pi$.

So $A - B = \dfrac{\pi}{2}$.

Question 18

a Area of $\Delta ABD = \dfrac{AD \times BD}{2}$

$\sin\theta = \dfrac{AD}{a}$ $\quad\quad$ $\cos\theta = \dfrac{BD}{a}$

$a\sin\theta = AD$ $\quad\quad$ $a\cos\theta = BD$

Area of $\Delta ABD = \dfrac{a\sin\theta \times a\cos\theta}{2}$

$= \dfrac{a^2 \sin\theta\cos\theta}{2}$

b Area of $\Delta ABC = \dfrac{1}{2} \times a \times a\sin 2\theta$ $\quad (\frac{1}{2}ab\sin C)$

$= \dfrac{a^2 \sin 2\theta}{2}$

Area of $\Delta ABC = 2 \times$ area of ΔABD

$= \dfrac{2a^2 \sin\theta\cos\theta}{2}$

$= a^2 \sin\theta\cos\theta$

So $\dfrac{a^2 \sin 2\theta}{2} = a^2 \sin\theta\cos\theta$

$a^2 \sin 2\theta = 2a^2 \sin\theta\cos\theta$

$\sin 2\theta = 2\sin\theta\cos\theta$

Question 19

a $\sin 2x = 2\sin^2 x$

$2\sin x\cos x = 2\sin^2 x$

$2\sin x\cos x - 2\sin^2 x = 0$

$2\sin x(\cos x - \sin x) = 0$

$2\sin x = 0$

Solutions $x = 0, \pi$ not in range.

OR $\cos x - \sin x = 0$

$\sin x = \cos x$

$\dfrac{\sin x}{\cos x} = 1$

$\tan x = 1$

So $x = \dfrac{\pi}{4}$ given $0 < x < \pi$.

b To prove that $\sin 2x > 2\sin^2 x$, first prove that $\sin 2x - 2\sin^2 x > 0$.

$\sin 2x - 2\sin^2 x$

$= 2\sin x \cos x - 2\sin^2 x$

$= 2\sin x(\cos x - \sin x)$ [*]

$= 2\sin x \cos x\left(\dfrac{\cos x - \sin x}{\cos x}\right)$ to create a $\tan x$ term

$= \sin 2x(1 - \tan x)$

Hint
You could also argue that from [*], that for $0 < x < \frac{\pi}{4}$, $\sin x > 0$ and $\cos x > \sin x$, so $2\sin x(\cos x - \sin x) > 0$.

Now for $0 < x < \dfrac{\pi}{4}$, $\sin 2x > 0$, $0 < \tan x < 1$, so $1 - \tan x > 0$, so $\sin 2x(1 - \tan x) > 0$.

$\therefore \sin 2x - 2\sin^2 x > 0$.

So $\sin 2x > 2\sin^2 x$.

Question 20

a

$$\text{LHS} = \sin^3\theta - \frac{3}{4}\sin\theta + \frac{\sin(3\theta)}{4}$$
$$= \frac{1}{4}[4\sin^3\theta - 3\sin\theta + \sin(3\theta)]$$

Now $\sin(3\theta) = \sin(2\theta + \theta)$

$= \sin 2\theta\cos\theta + \cos 2\theta\sin\theta$ $\quad(\sin(A + B) = \sin A\cos B + \cos A\sin B)$

$= 2\sin\theta\cos\theta\cos\theta + \cos 2\theta\sin\theta$ $\quad(\sin 2\theta = 2\sin\theta\cos\theta)$

$= 2\sin\theta\cos^2\theta + (\cos^2\theta - \sin^2\theta)\sin\theta$ $\quad(\cos 2\theta = \cos^2\theta - \sin^2\theta)$

$= 2\sin\theta\cos^2\theta + \sin\theta\cos^2\theta - \sin^3\theta$

$= 3\sin\theta\cos^2\theta - \sin^3\theta$

$= 3\sin\theta(1 - \sin^2\theta) - \sin^3\theta$ $\quad(\cos^2\theta = 1 - \sin^2\theta)$

$= 3\sin\theta - 3\sin^3\theta - \sin^3\theta$

$= 3\sin\theta - 4\sin^3\theta$

$$\text{So LHS} = \frac{1}{4}[4\sin^3\theta - 3\sin\theta + 3\sin\theta - 4\sin^3\theta]$$
$$= \frac{1}{4}[0]$$
$$= 0$$
$$= \text{RHS}$$

b

$$x^3 - 12x + 8 = 0$$
$$(4\sin\theta)^3 - 12(4\sin\theta) + 8 = 0$$
$$64\sin^3\theta - 48\sin\theta + 8 = 0$$
$$\sin^3\theta - \frac{48}{64}\sin\theta + \frac{8}{64} = 0$$
$$\sin^3\theta - \frac{3}{4}\sin\theta + \frac{1}{8} = 0$$

From part **a**: $\dfrac{1}{8} = \dfrac{\sin(3\theta)}{4}$

$\therefore \dfrac{1}{2} = \sin(3\theta)$

c $\sin 3\theta = \frac{1}{2}$ (Quadrants 1 and 2)

$$3\theta = \sin^{-1}\left(\frac{1}{2}\right)$$

$$3\theta = \frac{\pi}{6}, \pi - \frac{\pi}{6}, 2\pi + \frac{\pi}{6}, 3\pi - \frac{\pi}{6}, 4\pi + \frac{\pi}{6}, \ldots$$

$$3\theta = \frac{\pi}{6}, \frac{5\pi}{6}, \frac{13\pi}{6}, \frac{17\pi}{6}, \frac{25\pi}{6}, \ldots$$

$$\theta = \frac{\pi}{18}, \frac{5\pi}{18}, \frac{13\pi}{18}, \frac{17\pi}{18}, \frac{25\pi}{18}, \ldots$$

As $x = 4\sin\theta$: $4\sin\frac{\pi}{18}, 4\sin\frac{5\pi}{18}, 4\sin\frac{13\pi}{18}, \ldots$ are roots but are not all distinct.

3 distinct roots: $\alpha = 4\sin\frac{\pi}{18}$ (Quadrant 1)

$\beta = 4\sin\frac{5\pi}{18}$ (Quadrant 1)

$\gamma = 4\sin\frac{25\pi}{18}$ (Quadrant 3)

$$\begin{aligned}\alpha^2 + \beta^2 + \gamma^2 &= (\alpha + \beta + \gamma)^2 - 2(\alpha\beta + \alpha\gamma + \beta\gamma) \\ &= \left(-\frac{b}{a}\right)^2 - 2\left(\frac{c}{a}\right) \\ &= \left(-\frac{0}{1}\right)^2 - 2\left(\frac{-12}{1}\right) \\ &= 0 - 2(-12) \\ &= 24\end{aligned}$$

(from $x^3 + 0x^2 - 12x + 8 = 0$ in part **b**)

$$\begin{aligned}\alpha^2 + \beta^2 + \gamma^2 &= \left(4\sin\frac{\pi}{18}\right)^2 + \left(4\sin\frac{5\pi}{18}\right)^2 + \left(4\sin\frac{25\pi}{18}\right)^2 \\ &= 4^2\sin^2\frac{\pi}{18} + 4^2\sin^2\frac{5\pi}{18} + 4^2\sin^2\frac{25\pi}{18} \\ &= 16\left(\sin^2\frac{\pi}{18} + \sin^2\frac{5\pi}{18} + \sin^2\frac{25\pi}{18}\right)\end{aligned}$$

So $16\left(\sin^2\frac{\pi}{18} + \sin^2\frac{5\pi}{18} + \sin^2\frac{25\pi}{18}\right) = 24$

$$\sin^2\frac{\pi}{18} + \sin^2\frac{5\pi}{18} + \sin^2\frac{25\pi}{18} = \frac{24}{16}$$

$$\sin^2\frac{\pi}{18} + \sin^2\frac{5\pi}{18} + \sin^2\frac{25\pi}{18} = \frac{3}{2}$$

HSC exam topic grid (2011–2020)

This table shows the coverage of this topic in past HSC exams by question number. The past exams can be downloaded from the NESA website (www.educationstandards.nsw.edu.au) by selecting 'Year 11 – Year 12', 'HSC exam papers'. NESA marking feedback and guidelines can also be found there.

The new Mathematics Extension 1 course was first examined in 2020. For exams before 2020, select 'Year 11 – Year 12', 'Resources archive', 'HSC exam papers archive'.

	Auxiliary angle method	**Trigonometric equations using identities**	**Proving identities**
2011		5(a)(v)	5(a)(ii)
2012			
2013	12(a)	6	8
2014	**2**		
2015	11(d)		
2016		6	3
2017	4		
2018	11(c)	9	
2019	12(b)(i)–(ii)		6
2020 new course	**11(d)**		**14(b)**

Questions in **bold** can be found in this chapter.

WORKED SOLUTIONS

CHAPTER 4
FURTHER INTEGRATION

ME-C2 Further calculus skills 81
ME-C3 Applications of calculus 83
C3.1 Further area and volumes of solids of revolution 83

4

FURTHER INTEGRATION

Integration by substitution

- Let $u = \ldots$
- Given the substitution
- Choosing the substitution
- Trigonometric substitutions

Trigonometric integrals

- Integrating $\sin^2 x$ and $\cos^2 x$
- Trigonometric identities
- Products to sums formulas

Inverse trigonometric functions

- Differentiating inverse functions
- Differentiating inverse trigonometric functions
- Integrals involving inverse functions

Areas of integration

- Areas about the x-axis
- Areas about the y-axis

Volumes of solids of revolution

- Volumes about the x-axis
- Volumes about the y-axis

Glossary

A+ DIGITAL FLASHCARDS

Revise this topic's key terms and concepts by scanning the QR code or typing the URL into your browser.

https://get.ga/a-hsc-maths-ext-1

integrand

The function that is to be integrated.

integration by substitution

A method used for a composite or nested function in which a variable (usually u) is substituted for the inside or nested function and then differentiated. The integral can then be simplified into a more familiar form so that it can be integrated using the reverse chain rule.

inverse function

A function that performs the reverse of the original.

normal to a curve

The straight line that is perpendicular to a tangent to the curve at a specific point.

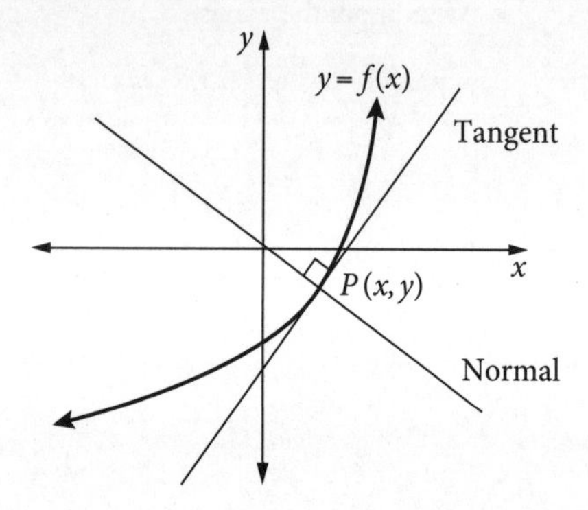

solid of revolution

A solid formed by rotating a region or a part of a curve around one of the coordinate axes. The formula for the solid formed when the area under the graph of $y = f(x)$ between $x = a$ and $x = b$ is rotated about the x-axis is:

$$V = \pi \int_a^b y^2\, dx$$

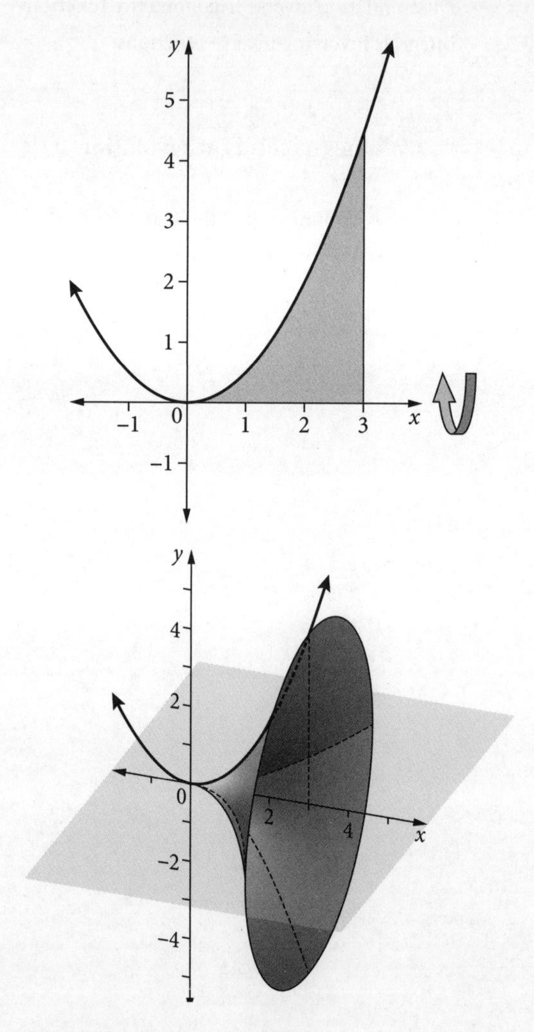

tangent to a curve

A straight line that 'touches' the graph of a function at a specific value of x. Its gradient measures the rate of change of the curve at that point.

Topic summary

Further calculus skills (ME-C2)

Integration by substitution: Steps

1. Choose the substitution 'Let $u = \ldots$' if not given in the question and then look for a part of the integral whose derivative is similar to another part of the integrand.
2. Find the derivative of the substitution.
3. Find the changed limits of integration if it is a definite integral.
4. Make substitutions to eliminate the original variable from the integral.
5. Integrate the new function.
6. Change the answer back to the original variable if it is an indefinite integral.

Choosing a substitution

u as a function of x

- Check for a function that appears inside of another function. (See Practice set 1, Question 16)
- Consider what the derivative of a part of the function will look like. (See Practice set 2, Question 18b)
- In a product of powers of sine and cosine ratios ($\sin^m x \cos^n x$):
  - If either m or n is odd, then choose the trigonometric ratio with the odd power and change all but one into the other ratio. ($\sin^2 x + \cos^2 x = 1$ will be required)
  - If both m and n are even, use the $\sin 2x$ formula first.
- In a product of powers of tangent and secant ratios ($\tan^m x \sec^n x$):
  - If the power of sec is even, choose $u = \tan x$.
  - If the power of sec is odd, choose $u = \sec x$. (Note that $\frac{d}{dx}\sec x = \sec x \tan x$)

x as a function of θ

- If the derivative of u is not apparent, a substitution into x can simplify the integrand. Choose a substitution that will simplify any square roots or complex sections of the integral.
- If the function contains powers of functions in the form $(a^2 - x^2)^n$, where n is fractional, consider using a trigonometric substitution $x = \sin \theta$.

Note: This type of substitution should always be given to you in Maths Extension 1 (see 2020 HSC Extension 1 exam, Question 13(a)) but may not be given at Maths Extension 2 level.

Hint
The more you practise these questions, the easier you will recognise the types of substitutions that will be appropriate for certain integrals. There will also be situations in which different substitutions are available but some will be easier than others. If you get stuck halfway through a problem, choose a different substitution as it may work better.

TOPIC SUMMARY

Integrating $\sin^2 x$ and $\cos^2 x$

Use these identities to integrate $\sin^2 x$ and $\cos^2 x$:

$$\sin^2 x = \frac{1}{2}(1 - \cos 2x)$$

$$\cos^2 x = \frac{1}{2}(1 + \cos 2x)$$

More generalised formulas appear on the HSC exam reference sheet (also at the back of this book):

$$\sin^2 nx = \frac{1}{2}(1 - \cos 2nx)$$

$$\cos^2 nx = \frac{1}{2}(1 + \cos 2nx)$$

Further trigonometric derivatives and integrals

These can be proven by differentiating the reciprocals of sine, cosine and tangent.

$$\frac{d}{dx}\sec x = \sec x \tan x$$

$$\frac{d}{dx}\operatorname{cosec} x = -\operatorname{cosec} x \cot x$$

$$\frac{d}{dx}\cot x = -\operatorname{cosec}^2 x$$

Hint

These formulas are NOT on the HSC exam reference sheet but you should notice a pattern between the sec and cosec results, and the tan and cotan results: change to co- and make negative.

These integrals can be derived from the derivatives.

$$\int \sec x \tan x \, dx = \sec x + c$$

$$\int \operatorname{cosec} x \cot x \, dx = -\operatorname{cosec} x + c$$

$$\int \operatorname{cosec}^2 x \, dx = -\cot x + c$$

Products to sums formulas

When integrating a function containing the product of 2 trigonometric ratios where the angles are different, use the products to sums formulas. Split the product into the sum or difference of sine and cosine that can then be easily integrated. These formulas appear on the HSC exam reference sheet.

$$\cos A \cos B = \frac{1}{2}[\cos(A - B) + \cos(A + B)]$$

$$\sin A \sin B = \frac{1}{2}[\cos(A - B) - \cos(A + B)]$$

$$\sin A \cos B = \frac{1}{2}[\sin(A + B) + \sin(A - B)]$$

$$\cos A \sin B = \frac{1}{2}[\sin(A + B) - \sin(A - B)]$$

Inverse functions

The derivative of the function can be determined by using the inverse:

$$\frac{dy}{dx} = \frac{1}{\frac{dx}{dy}}$$

Inverse trigonometric functions

Derivatives

$$\frac{d}{dx}\sin^{-1} f(x) = \frac{f'(x)}{\sqrt{1-[f(x)]^2}}$$

$$\frac{d}{dx}\cos^{-1} f(x) = \frac{-f'(x)}{\sqrt{1-[f(x)]^2}}$$

$$\frac{d}{dx}\tan^{-1} f(x) = \frac{f'(x)}{1+[f(x)]^2}$$

Integrals

$$\int \frac{f'(x)}{\sqrt{a^2-[f(x)]^2}}\,dx = \sin^{-1}\frac{f(x)}{a} + c$$

$$\int \frac{f'(x)}{a^2+[f(x)]^2}\,dx = \frac{1}{a}\tan^{-1}\frac{f(x)}{a} + c$$

Hint
These formulas are on the HSC exam reference sheet.

Note: The matching integral that gives $\cos^{-1} x$ is not given because it is the negative of the integral that gives $\sin^{-1} x$. Use either when integrating.

Hint
The generalised integrals showing $f(x)$ on the HSC exam reference sheet can seem cumbersome. You may want to learn them in a simpler form and then apply the reverse chain rule or integration by substitution. For example:

Learn $\int \frac{1}{\sqrt{a^2-x^2}}\,dx = \sin^{-1}\left(\frac{x}{a}\right) + c$ instead of $\int \frac{f'(x)}{\sqrt{a^2-[f(x)]^2}}\,dx = \sin^{-1}\frac{f(x)}{a} + c$.

Use $\int \frac{1}{a^2+x^2} = \frac{1}{a}\tan^{-1}\left(\frac{x}{a}\right) + c$ instead of $\int \frac{f'(x)}{a^2+[f(x)]^2}\,dx = \frac{1}{a}\tan^{-1}\frac{f(x)}{a} + c$.

Applications of calculus (ME-C3)

C3.1 Further area and volumes of solids of revolution

Areas about the x-axis

Areas about the x-axis would have been encountered when you first learnt integration in Maths Advanced.

$$A = \int_a^b y\,dx$$

The limits a and b are the x-coordinates.

The shaded area (in units^2) bounded by the curve $y = x^2$ and the x-axis between $x = 0$ and $x = 2$ can be written as:

$$A_1 = \int_0^2 x^2\,dx$$

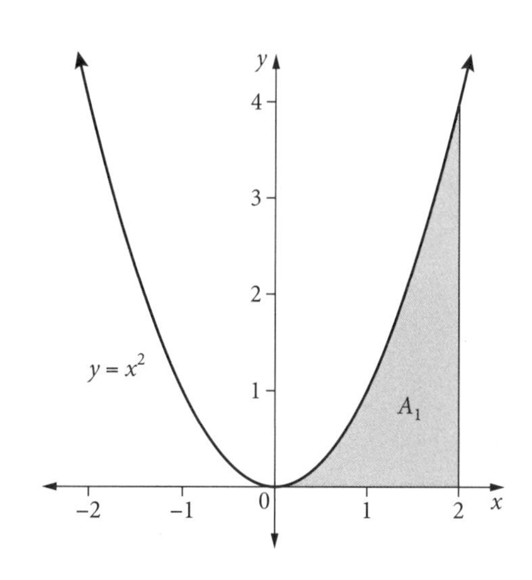

Areas about the y-axis

This formula is not strictly part of the course but handy to know for volumes about the y-axis below.

Areas about the y-axis changes the integral to:

$$A = \int_a^b x\, dy$$

The limits a and b are the y-coordinates.

For the shaded area, since

$$y = x^2$$
$$x = \pm\sqrt{y}$$

Take the positive square root as $x \geq 0$.

The shaded area (in units²) bounded by the curve $y = x^2$ and the y-axis between $y = 0$ and $y = 4$ can be written as:

$$A_2 = \int_0^4 \sqrt{y}\, dy$$

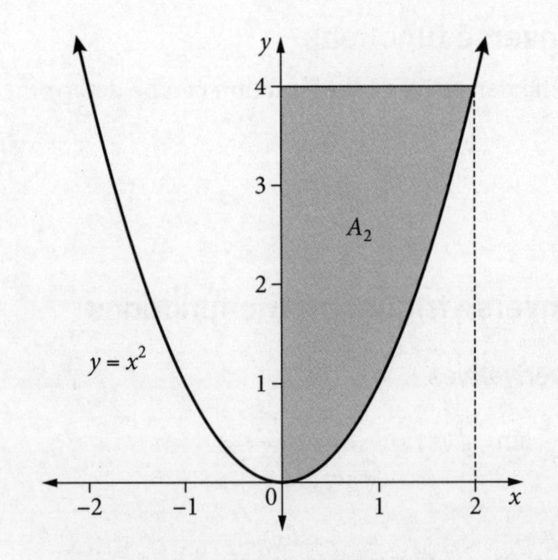

Hint

The shaded area can also be found by calculating the area under the curve and subtracting it from a 2 × 4 rectangle.

$$A_2 = 2 \times 4 - \int_0^2 x^2 dx$$

Volumes of solids of revolution: steps

1. Draw a diagram of the region if one is not given.
2. Find the endpoints of the region to be rotated. Remember that the endpoints must be on the axis of rotation.
3. Write out the required integral.
4. Solve the integral for the solution.

Hint

When approaching these problems, sketch the graph and shade in the area to be rotated. It also helps to draw in an arrow indicating the direction of rotation.

Volumes rotated about the x-axis

$$V = \pi\int_a^b y^2\,dx$$

The area (in units^2) bounded by the curve $y = \frac{1}{2}x^2$ and the x-axis between $x = 0$ and $x = 3$ is given by:

$$A = \int_0^3 \frac{1}{2}x^2 dx$$

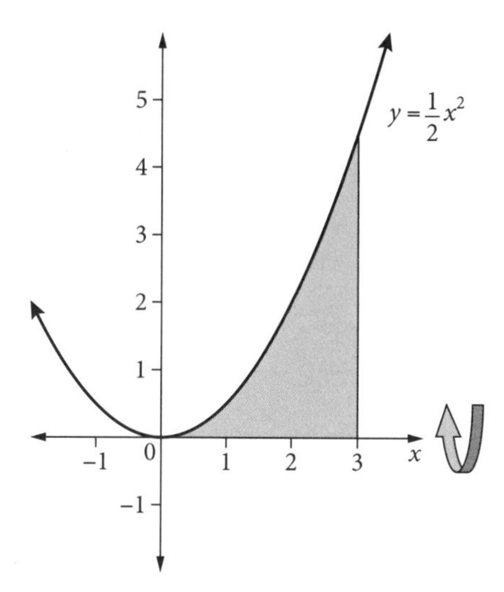

The volume (in units^3) bounded by the curve $y = \frac{1}{2}x^2$ and the x-axis between $x = 0$ and $x = 3$ is given by:

$$V = \pi\int_0^3 \left(\frac{1}{2}x^2\right)^2 dx$$

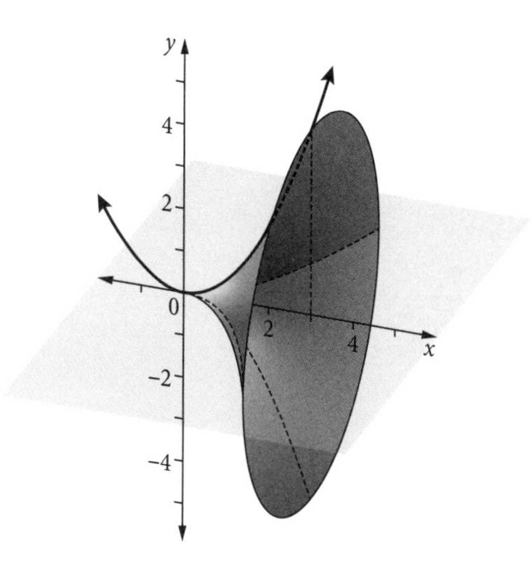

Volumes rotated about the y-axis

$$V = \pi\int_a^b x^2\,dy$$

The area (in units^2) bounded by the curve $y = \frac{1}{2}x^2$ and the y-axis between $y = 0$ and $y = 4$ is given by:

$$A = \int_0^4 \sqrt{2y}\,dy$$

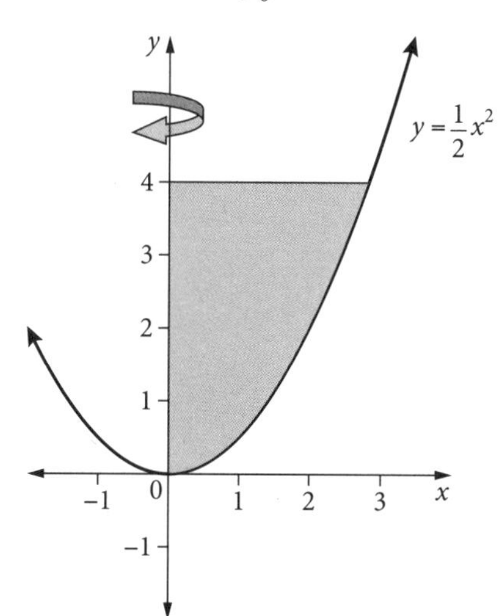

The volume (in units^3) bounded by the curve $y = \frac{1}{2}x^2$ and the y-axis between $y = 0$ and $y = 4$ is given by:

$$V = \pi\int_0^4 2y\,dy$$

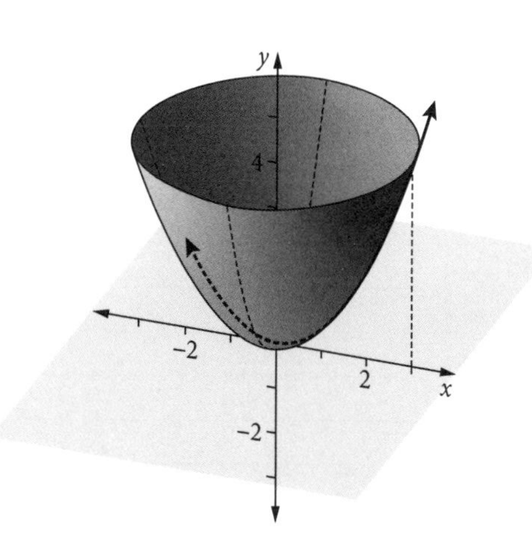

Note: If rotating a region made by a pair of curves, volumes may need to be added or subtracted.

Practice sets tracking grid

Maths is all about repetition, meaning do, do and do again! Each question in the following practice sets, especially the struggle questions (different for everybody!), should be completed at least 3 times correctly. Below is a tracking grid to record your question attempts: ✓ if you answered correctly,
✗ if you didn't.

PRACTICE SET 1: Multiple-choice questions

Question	1st attempt	2nd attempt	3rd attempt	4th attempt	5th attempt
1					
2					
3					
4					
5					
6					
7					
8					
9					
10					
11					
12					
13					
14					
15					
16					
17					
18					
19					
20					

PRACTICE SET 2: Short-answer questions

Question	1st attempt	2nd attempt	3rd attempt	4th attempt	5th attempt
1					
2					
3					
4					
5					
6					
7					
8					
9					
10					
11					
12					
13					
14					
15					
16					
17					
18					
19					
20					
21					

9780170459242

Practice set 1

Multiple-choice questions

Solutions start on page 94.

Question 1

Find $\int \frac{3}{9+x^2}\,dx$.

A $3\tan^{-1}x + c$ **B** $3\tan^{-1}\left(\frac{x}{3}\right) + c$ **C** $\tan^{-1}x + c$ **D** $\tan^{-1}\left(\frac{x}{3}\right) + c$

Question 2

Find $\int \frac{x}{\sqrt{x^2-2}}\,dx$.

A $\ln\left(\sqrt{x^2-2}\right) + c$ **B** $\sqrt{x^2-2} + c$ **C** $2\sqrt{x^2-2} + c$ **D** $\frac{1}{2}\sqrt{x^2-2} + c$

Question 3

What is the gradient of the tangent of $y = \cos^{-1}\left(\frac{2x}{5}\right)$ when $x = 2$?

A $-\frac{2}{3}$ **B** $-\frac{1}{3}$ **C** $\frac{1}{3}$ **D** $\frac{2}{3}$

Question 4

What is the exact value of $\int_0^{\frac{\pi}{8}} \cos^2 x\,dx$?

A $\frac{\pi - 2\sqrt{2}}{16}$ **B** $\frac{\pi - 2\sqrt{2}}{8}$ **C** $\frac{\pi + 2\sqrt{2}}{16}$ **D** $\frac{\pi + 2\sqrt{2}}{8}$

Question 5

What is the exact value of $\int_{-\frac{\pi}{4}}^{\frac{\pi}{4}} \sec^2 x + \cos^2 x\,dx$?

A $\frac{10+\pi}{8}$ **B** $\frac{10+\pi}{4}$ **C** $\frac{6+\pi}{8}$ **D** $\frac{6+\pi}{4}$

Question 6

Using the substitution $u = 4 - x$, which of the following is equivalent to $\int_0^1 \frac{x}{\sqrt{4-x}}\,dx$?

A $\int_3^4 \frac{4-u}{\sqrt{u}}\,du$ **B** $\int_0^1 \frac{4-u}{\sqrt{u}}\,du$ **C** $\int_3^4 \frac{u-4}{\sqrt{u}}\,du$ **D** $\int_0^1 \frac{u-4}{\sqrt{u}}\,du$

Question 7

Find $\int \sin 4x \cos 4x\,dx$.

A $-\frac{1}{8}\cos 8x + c$ **B** $\frac{1}{8}\cos 8x + c$ **C** $-\frac{1}{16}\cos 8x + c$ **D** $\frac{1}{16}\cos 8x + c$

Question 8

The region between $y = 2\sin x$ and the x-axis between $x = 0$ and $x = \pi$ is rotated about the x-axis.

What is the volume of the solid formed?

A 4 units^3 **B** 2π units^3 **C** 4π units^3 **D** $2\pi^2$ units^3

Question 9

Which integral can be used to find the volume of the solid created by rotating the shaded region in the graph about the y-axis?

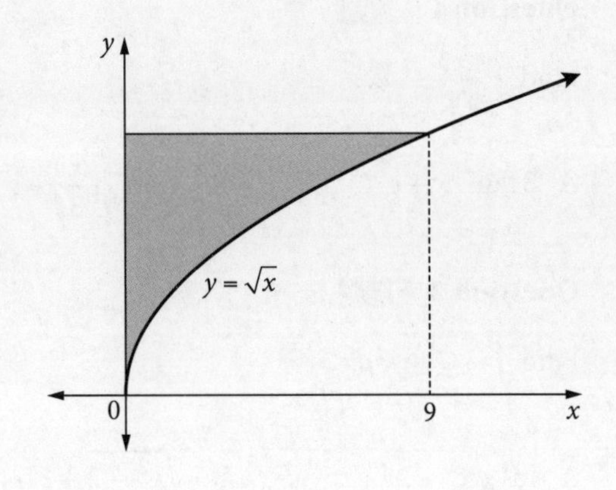

A $V = \pi\int_0^3 y^4\, dy$

B $V = \pi\int_0^9 y^4\, dy$

C $V = \pi\int_0^3 y^2\, dy$

D $V = \pi\int_0^9 y^2\, dy$

Question 10

What is the equation of the normal to $y = 3\tan^{-1}\left(\frac{x-2}{2}\right)$ at the point $\left(4, \frac{3\pi}{4}\right)$?

A $x - 8y - 6\pi - 4 = 0$

B $3x - 8y - 6\pi - 12 = 0$

C $8x + 4y - 3\pi - 32 = 0$

D $32x + 12y - 9\pi - 128 = 0$

Question 11

Find $\frac{d}{dx}\left(x^2\cos^{-1}x\right)$.

A $-\frac{2x}{\sqrt{1-x^2}}$

B $2x\cos^{-1}x - \frac{x^2}{\sqrt{1-x^2}}$

C $2x\cos^{-1}x + \frac{x^2}{\sqrt{1-x^2}}$

D $\frac{2x}{\sqrt{1-x^2}}$

Question 12

For what value of p is $\int_0^p \frac{2}{\sqrt{4-x^2}}\, dx = \frac{\pi}{2}$?

A $\frac{1}{2}$

B $\frac{1}{\sqrt{2}}$

C $\sqrt{2}$

D 2

Question 13

Find $\int \cos 4x \sin x\, dx$.

A $\frac{1}{3}\cos 3x - \frac{1}{5}\cos 5x + c$

B $\frac{1}{6}\cos 3x - \frac{1}{10}\cos 5x + c$

C $-\frac{1}{4}\sin 4x\cos x + c$

D $\frac{1}{4}\sin 4x\cos x + c$

Question 14

The region bounded by $y = 7x - 2x^2$ and the x-axis is rotated about the x-axis.

What is the volume of the solid formed?

A 2401 units^3 **B** $\frac{343\pi}{24}$ units^3 **C** $\frac{343\pi}{6}$ units^3 **D** $\frac{16\,807\pi}{240}$ units^3

Question 15

The region enclosed by $y = \ln x + 1$, the line $y = x$, and the x-axis is rotated about the y-axis.

What is the volume of the solid of revolution formed?

A $\frac{\pi}{6}(5 - e^{-2})$ **B** $\frac{\pi}{6}(5 - 3e^{-2})$ **C** $\frac{\pi}{6}(1 - e^{-2})$ **D** $\frac{\pi}{6}(1 - 3e^{-2})$

Question 16

Use the substitution $u = x^2 + 6x - 5$ to find $\int (x + 3)\sqrt[3]{x^2 + 6x - 5}\,dx$.

A $\frac{3}{8}(x^2 + 6x - 5)^{\frac{4}{3}} + c$ **B** $\frac{1}{3}(x^2 + 6x - 5)^{\frac{3}{2}} + c$ **C** $\frac{3}{4}(x^2 + 6x - 5)^{\frac{3}{4}} + c$ **D** $\sqrt[3]{x^2 + 6x - 5} + c$

Question 17

Evaluate $\int_0^1 \frac{x^2}{\sqrt{4 - x^2}}\,dx$ using the substitution $x = 2\sin\theta$.

A $\frac{\sqrt{3}}{2} - \frac{\pi}{3}$ **B** $\frac{\pi}{3} + \frac{\sqrt{3}}{2}$ **C** $\frac{\pi}{3} - \frac{\sqrt{3}}{2}$ **D** $2 - \sin 2$

Question 18

Find $\int \frac{e^{-2x}}{e^{-x} + 1}\,dx$ using the substitution $u = e^{-x} + 1$.

A $\ln(e^{-x} + 1) + c$

B $(e^{-x} + 1) - \ln(e^{-x} + 1) + c$

C $\ln(e^{-x} + 1) - (e^{-x} + 1) + c$

D $\frac{e^{-2x}}{2(e^{-x} + 1)} + c$

Question 19

Find $f(x)$ if $f'(x) = \frac{1}{\sqrt{3 - 2x - x^2}}$ and $f(1) = \frac{\pi}{4}$.

A $f(x) = \sin^{-1}\left(\frac{x+1}{2}\right)$

B $f(x) = \sin^{-1}\left(\frac{x+1}{2}\right) - \frac{\pi}{4}$

C $f(x) = \sin^{-1}\left(\frac{x+1}{4}\right)$

D $f(x) = \sin^{-1}\left(\frac{x+1}{4}\right) + \frac{\pi}{12}$

Question 20

The region between $y = \tan^{-1} x$ and the y-axis between $y = 0$ and $y = \frac{\pi}{3}$ is rotated about the y-axis.

What is the volume of the solid of revolution formed?

A 3 units^3 **B** $\pi \ln 2$ units^3 **C** $\frac{\pi}{3}(3\sqrt{3} - \pi)$ units^3 **D** $\frac{\pi}{3}(3\sqrt{3} + \pi)$ units^3

Practice set 2

Short-answer questions

Solutions start on page 98.

Question 1 (10 marks)

Differentiate each function.

a $y = \sin^{-1} 2x$ — 2 marks

b $y = \ln(\sin^{-1} 3x)$ — 2 marks

c $y = e^{\cos^{-1} x}$ — 2 marks

d $y = (\tan^{-1} x)^3$ — 2 marks

e $y = \tan^{-1}\left(\frac{2}{x}\right)$ — 2 marks

Question 2 (4 marks)

Evaluate each definite integral.

a $\int_0^1 \frac{1}{16 + x^2}\, dx$ — 2 marks

b $\int_{-\sqrt{3}}^0 \frac{1}{\sqrt{4 - x^2}}\, dx$ — 2 marks

Question 3 (4 marks)

Find each integral.

a $\int \frac{1}{16 + 9x^2}\, dx$ — 2 marks

b $\int \frac{1}{4x^2 + 4x + 2}\, dx$ — 2 marks

Question 4 (3 marks)

a Show that $7 + 12x - 4x^2 = 16 - (2x - 3)^2$. — 1 mark

b Hence, find $\int_0^2 \frac{1}{\sqrt{7 + 12x - 4x^2}}\, dx$. Leave your answer in exact form. — 2 marks

Question 5 (3 marks)

Use a products to sums formula to evaluate $\int_0^{\frac{\pi}{2}} \cos 5x \cos 2x\, dx$. — 3 marks

Question 6 (3 marks)

Find the gradient of the tangent to the curve $y = \sin^{-1} 3x$ at the point where $x = \frac{\sqrt{3}}{6}$. — 3 marks

Question 7 (3 marks)

The shaded region bounded by the curve $y = \sqrt{x-1}$, the coordinate axes and the line $y = 2$ shown below is rotated about the y-axis.

Find the volume of the solid generated. 3 marks

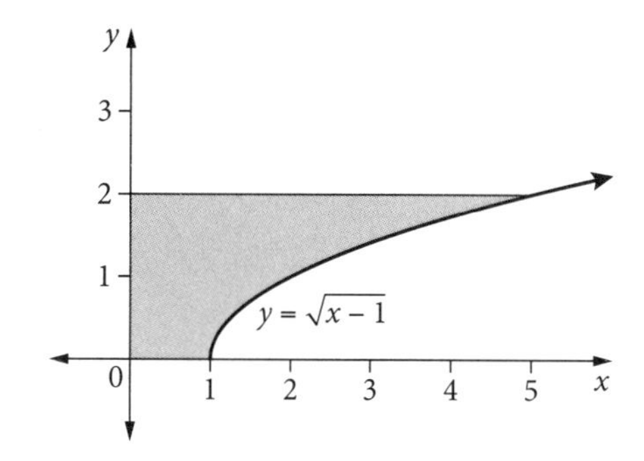

Question 8 (6 marks)

Find each integral.

a $\int \sin^3\theta \cos^5\theta \, d\theta$ 2 marks

b $\int \tan\theta \sec^3\theta \, d\theta$ 2 marks

c $\int \cos^4 3\theta \, d\theta$ 2 marks

Question 9 (6 marks)

Evaluate each definite integral.

a $\int_{\frac{\pi}{4}}^{\frac{\pi}{2}} \cot^2 x \, dx$ 2 marks

b $\int_0^{\frac{\pi}{4}} \tan^2\theta \sec^4\theta \, d\theta$ 2 marks

c $\int_{\frac{\pi}{2}}^{2\pi} 4\sin^2\left(\frac{x}{4}\right)\cos^2\left(\frac{x}{4}\right) dx$ 2 marks

Question 10 (3 marks)

a Show that $4\sin 2x \sin 4x \sin 6x = \sin 4x + \sin 8x - \sin 12x$. 2 marks

b Hence, find $\int \sin 2x \sin 4x \sin 6x \, dx$. 1 mark

Question 11 (4 marks)

a Show $\dfrac{2x^2 - 2x - 5}{x^2 + 1} = 2 - \dfrac{2x + 7}{x^2 + 1}$. 1 mark

b Hence, evaluate $\int_1^6 \dfrac{2x^2 - 2x - 5}{x^2 + 1} dx$. Write your answer to two decimal places. 3 marks

9780170459242

Question 12 (9 marks)

Find each integral.

a $\int \frac{\tan^{-1} x}{1+x^2}\,dx$ (Use $u = \tan^{-1} x$) 3 marks

b $\int \frac{x+1}{\sqrt{x^2+2x}}\,dx$ (Use $u = x^2 + 2x$) 3 marks

c $\int_0^1 x\sqrt{1-x}\,dx$ (Use $u = 1 - x$) 3 marks

Question 13 (3 marks)

Given that $\frac{d^2y}{dx^2} = \frac{x}{\sqrt{(9-x^2)^3}}$, find y in terms of x if $\frac{dy}{dx} = \frac{1}{3}$ and $y = \pi$ when $x = 0$. 3 marks

Question 14 (3 marks)

a Show that $\frac{d}{dx}\sin(\cos^{-1} x) = -\frac{x}{\sqrt{1-x^2}}$ for $[-1, 1]$. 1 mark

b Hence, or otherwise, show that $\sin(\cos^{-1} x) = \sqrt{1-x^2}$. 2 marks

Question 15 (7 marks)

a Graph the curve $xy = 4$ and the line $x + y = 5$ on the same number plane, showing key features of each graph and points of intersection. 2 marks

b Find the area bounded by the two graphs. 2 marks

c Find the volume of the solid formed by rotating this region about the x-axis. 3 marks

Question 16 (4 marks)

a Differentiate $y = x\sin^{-1} x$. 1 mark

b Hence, find $\int \sin^{-1} x\,dx$. 3 marks

Question 17 (9 marks)

Find each integral using the substitution provided.

a $\int (1-x^2)^{\frac{3}{2}}\,dx$ (Use $x = \sin\theta$) 3 marks

b $\int \frac{1}{(16+x^2)^{\frac{3}{2}}}\,dx$ (Use $x = 4\tan\theta$) 3 marks

c $\int_{-1}^{0} \frac{x+1}{\sqrt{1-x}}\,dx$ (Use $u = \sqrt{1-x}$) 3 marks

Question 18 (6 marks)

Integrate each expression. You will need to choose an appropriate substitution.

a $\int x(1-x^2)^4\,dx$ 2 marks

b $\int \frac{1}{x\ln x}\,dx$ 2 marks

c $\int \frac{x}{4+9x^4}\,dx$ 2 marks

Question 19 (6 marks)

A region is bounded by the curve $y = \sqrt{x^3 - x^2 - 8x + 12}$ and the positive x- and y-axes.

a Given that the polynomial $f(x) = x^3 - x^2 - 8x + 12$ has integer roots, factorise the polynomial fully and sketch the graph of the curve with the region shaded. 3 marks

b This region is rotated about the x-axis. Find the volume of the solid of revolution. 3 marks

Question 20 (3 marks)

The semicircle $y = \sqrt{r^2 - x^2}$ is rotated about the x-axis.

Prove that the volume of the sphere generated is $\frac{4}{3}\pi r^3$. 3 marks

Question 21 (6 marks)

The bowl section of a wine glass (not the stem) is formed by rotating the curve $y = 3\sin\left(\frac{x}{3}\right)$ about the x-axis between $x = 0$ and $x = 2\pi$, where x and y are measured in centimetres.

a Find, in exact form in cubic centimetres, the capacity of the wine glass. 2 marks

b Find, in cubic centimetres, an expression for the volume of the wine in the glass when it is h cm high. 2 marks

c If the wine glass is filled to half its height, find in exact form the ratio of the volume of wine to air in the glass. 2 marks

Practice set 1

Worked solutions

1 D

$$\int \frac{3}{9+x^2}\,dx$$
$$= 3\int \frac{1}{9+x^2}\,dx$$
$$= \frac{3}{3}\tan^{-1}\left(\frac{x}{3}\right) + c$$
$$= \tan^{-1}\left(\frac{x}{3}\right) + c$$

2 B

$$\int \frac{x}{\sqrt{x^2-2}}\,dx$$
$$= \frac{1}{2}\int 2x(x^2-2)^{-\frac{1}{2}}\,dx$$
$$= \frac{1}{2}\frac{(x^2-2)^{\frac{1}{2}}}{\frac{1}{2}} + c$$
$$= (x^2-2)^{\frac{1}{2}} + c$$
$$= \sqrt{x^2-2} + c$$

3 A

$$\frac{dy}{dx} = -\frac{1}{\sqrt{5^2-(2x)^2}} \times 2$$
$$= -\frac{2}{\sqrt{25-4x^2}}$$

When $x = 2$:

$$\frac{dy}{dx} = -\frac{2}{\sqrt{25-4\times 2^2}}$$
$$= -\frac{2}{\sqrt{9}}$$
$$= -\frac{2}{3}$$

4 C

$$\int_0^{\frac{\pi}{8}} \cos^2 x\,dx$$
$$= \frac{1}{2}\int_0^{\frac{\pi}{8}} 1 + \cos 2x\,dx$$
$$= \frac{1}{2}\left[x + \frac{1}{2}\sin 2x\right]_0^{\frac{\pi}{8}}$$
$$= \frac{1}{2}\left[\left(\frac{\pi}{8} + \frac{1}{2}\sin\frac{\pi}{4}\right) - \left(0 + \frac{1}{2}\sin 0\right)\right]$$
$$= \frac{1}{2}\left(\frac{\pi}{8} + \frac{\sqrt{2}}{4} - 0\right)$$
$$= \frac{\pi + 2\sqrt{2}}{16}$$

5 B

$$\int_{-\frac{\pi}{4}}^{\frac{\pi}{4}} \sec^2 x + \cos^2 x\,dx$$
$$= \int_{-\frac{\pi}{4}}^{\frac{\pi}{4}} \sec^2 x + \frac{1}{2}(1 + \cos 2x)\,dx$$
$$= \left[\tan x + \frac{1}{2}\left(x + \frac{1}{2}\sin 2x\right)\right]_{-\frac{\pi}{4}}^{\frac{\pi}{4}}$$
$$= \left(\tan\frac{\pi}{4} + \frac{1}{2}\left(\frac{\pi}{4} + \frac{1}{2}\sin\frac{\pi}{2}\right)\right) - \left(\tan\frac{-\pi}{4} + \frac{1}{2}\left(\frac{-\pi}{4} + \frac{1}{2}\sin\frac{-\pi}{2}\right)\right)$$
$$= \left(1 + \frac{1}{2}\left(\frac{\pi}{4} + \frac{1}{2}\right)\right) - \left(-1 + \frac{1}{2}\left(-\frac{\pi}{4} - \frac{1}{2}\right)\right)$$
$$= 1 + \frac{\pi}{8} + \frac{1}{4} - \left(-1 - \frac{\pi}{8} - \frac{1}{4}\right)$$
$$= 1 + \frac{\pi}{8} + \frac{1}{4} + 1 + \frac{\pi}{8} + \frac{1}{4}$$
$$= \frac{\pi}{4} + \frac{5}{2}$$
$$= \frac{\pi + 10}{4}$$

6 A

$u = 4 - x$

$x = 4 - u$

$dx = -du$

$x = 0, u = 4$

$x = 1, u = 3$

$$\int_0^1 \frac{x}{\sqrt{4-x}}\,dx = \int_4^3 \frac{4-u}{\sqrt{u}} \times -du$$

$$= \int_3^4 \frac{4-u}{\sqrt{u}}\,du$$

7 C

$$\int \sin 4x \cos 4x \, dx$$

$$= \frac{1}{2}\int \sin 8x \, dx$$

$$= -\frac{1}{16}\cos 8x + c$$

8 D

$$V = \pi \int y^2 \, dx$$

$$= \pi \int_0^{\pi} (2\sin x)^2 \, dx$$

$$= \pi \int_0^{\pi} 4\sin^2 x \, dx$$

$$= \pi \int_0^{\pi} 2(1 - \cos 2x)\, dx$$

$$= 2\pi \left[x - \frac{1}{2}\sin 2x\right]_0^{\pi}$$

$$= 2\pi\left(\pi - \frac{1}{2}\sin 2\pi - 0\right)$$

$$= 2\pi^2 \text{ units}^3$$

9 A

$x = 9, y = 3$

$y = \sqrt{x}, x = y^2$

$$V = \int \pi x^2 \, dy$$

$$= \int_0^3 \pi (y^2)^2 \, dy$$

$$= \int_0^3 \pi y^4 \, dy$$

10 D

$$y = 3\tan^{-1}\left(\frac{x-2}{2}\right)$$

$$\frac{dy}{dx} = 3 \times \frac{1}{4 + (x-2)^2}$$

When $x = 4$:

$$\frac{dy}{dx} = \frac{3}{4 + (4-2)^2}$$

$$= \frac{3}{8}$$

$$m_{\text{norm}} = -\frac{8}{3}$$

$$y - \frac{3\pi}{4} = -\frac{8}{3}(x - 4)$$

$$12y - 9\pi = -32x + 128$$

So $32x + 12y - 9\pi - 128 = 0$.

11 B

$u = x^2 \qquad u' = 2x$

$v = \cos^{-1} x \qquad v' = -\frac{1}{\sqrt{1-x^2}}$

$$\frac{d}{dx} x^2 \cos^{-1} x$$

$$= 2x\cos^{-1} x + x^2 \times \left(-\frac{1}{\sqrt{1-x^2}}\right)$$

$$= 2x\cos^{-1} x - \frac{x^2}{\sqrt{1-x^2}}$$

12 C

$$\int_0^p \frac{2}{\sqrt{4-x^2}}\,dx = \frac{\pi}{2}$$

$$\left[2\sin^{-1}\frac{x}{2}\right]_0^p = \frac{\pi}{2}$$

$$2\sin^{-1}\left(\frac{p}{2}\right) - 2\sin^{-1} 0 = \frac{\pi}{2}$$

$$2\sin^{-1}\left(\frac{p}{2}\right) = \frac{\pi}{2}$$

$$\sin^{-1}\left(\frac{p}{2}\right) = \frac{\pi}{4}$$

$$\frac{p}{2} = \sin\frac{\pi}{4}$$

$$p = 2\sin\frac{\pi}{4}$$

$$= 2 \times \frac{1}{\sqrt{2}}$$

$$= \sqrt{2}$$

13 B

$$\int \cos 4x \sin x \, dx$$
$$= \frac{1}{2}\int \sin(4x + x) - \sin(4x - x)\, dx$$
$$= \frac{1}{2}\int \sin 5x - \sin 3x \, dx$$
$$= \frac{1}{2}\left(-\frac{1}{5}\cos 5x + \frac{1}{3}\cos 3x\right) + c$$
$$= \frac{1}{6}\cos 3x - \frac{1}{10}\cos 5x + c$$

14 D

Determine roots of equation to find limits of integral.

$$7x - 2x^2 = 0$$
$$x(7 - 2x) = 0$$
$$x = 0, x = \frac{7}{2}$$

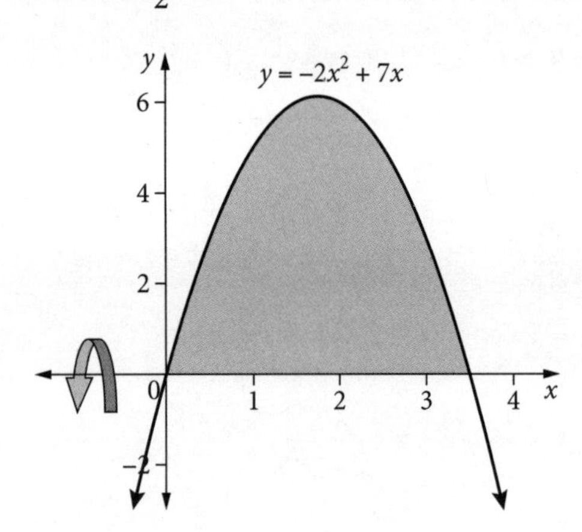

Calculate the volume of the solid.

$$V = \int \pi y^2 \, dx$$
$$= \pi\int_0^{\frac{7}{2}}\left(7x - 2x^2\right)^2 dx$$
$$= \pi\int_0^{\frac{7}{2}}(49x^2 - 28x^3 + 4x^4)\, dx$$
$$= \pi\left[\frac{49x^3}{3} - \frac{28x^4}{4} + \frac{4x^5}{5}\right]_0^{3.5}$$
$$= \pi\left(\frac{49 \times 3.5^3}{3} - 7 \times (3.5)^4 + \frac{4 \times 3.5^5}{5} - (0 - 0 + 0)\right)$$
$$= \pi \times \frac{16\,807}{240}$$
$$= \frac{16\,807\pi}{240} \text{ units}^3$$

15 D

Point of intersection:

$$\ln x + 1 = x$$
$$\ln x = x - 1$$
$$x = e^{x-1}$$
$$x = 1 \text{ (by inspection)}$$

$y = 1$, so $(1, 1)$.

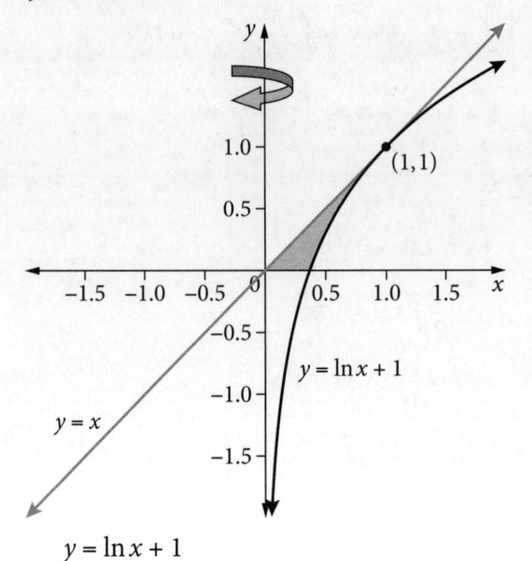

$$y = \ln x + 1$$
$$\ln x = y - 1$$
$$x = e^{y-1}$$

$$V = \pi\int_0^1 (e^{y-1})^2 \, dy - \pi\int_0^1 (y)^2 \, dy$$
$$= \pi\int_0^1 e^{2y-2} \, dy - \pi\int_0^1 y^2 \, dy$$
$$= \pi\left[\frac{1}{2}e^{2y-2}\right]_0^1 - \pi\left[\frac{y^3}{3}\right]_0^1$$
$$= \frac{\pi}{2}(e^{2-2} - e^{0-2}) - \frac{\pi}{3}(1^3 - 0^3)$$
$$= \frac{\pi}{2} - \frac{\pi e^{-2}}{2} - \frac{\pi}{3}$$
$$= \frac{\pi}{6}(1 - 3e^{-2}) \text{ units}^3$$

16 A

$$u = x^2 + 6x - 5 \qquad du = (2x + 6)\, dx$$
$$\frac{1}{2}\, du = (x + 3)\, dx$$

$$\int (x + 3)\sqrt[3]{x^2 + 6x - 5}\, dx$$
$$= \int \sqrt[3]{u} \times \frac{1}{2}\, du$$
$$= \frac{1}{2}\int u^{\frac{1}{3}}\, du$$
$$= \frac{1}{2} \times \frac{3}{4}u^{\frac{4}{3}} + c$$
$$= \frac{3}{8}u^{\frac{4}{3}} + c$$
$$= \frac{3}{8}(x^2 + 6x - 5)^{\frac{4}{3}} + c$$

17 C

$$\int_0^1 \frac{x^2}{\sqrt{4-x^2}}\,dx$$

$x = 2\sin\theta$

$dx = 2\cos\theta\, d\theta$

$x = 0,\ \theta = 0$

$x = 1,\ \theta = \frac{\pi}{6}$

$$= \int_0^{\frac{\pi}{6}} \frac{(2\sin\theta)^2}{\sqrt{4-(2\sin\theta)^2}} \times 2\cos\theta\, d\theta$$

$$= \int_0^{\frac{\pi}{6}} \frac{8\sin^2\theta\cos\theta}{\sqrt{4(1-\sin^2\theta)}}\, d\theta$$

$$= \int_0^{\frac{\pi}{6}} \frac{8\sin^2\theta\cos\theta}{2\cos\theta}\, d\theta$$

$$= \int_0^{\frac{\pi}{6}} 4\sin^2\theta\, d\theta$$

$$= 2\int_0^{\frac{\pi}{6}} 1 - \cos 2\theta\, d\theta$$

$$= 2\left[\theta - \frac{1}{2}\sin 2\theta\right]_0^{\frac{\pi}{6}}$$

$$= 2\left(\frac{\pi}{6} - \frac{1}{2}\sin\frac{\pi}{3} - (0-0)\right)$$

$$= \frac{\pi}{3} - \frac{\sqrt{3}}{2}$$

18 C

$$\int \frac{e^{-2x}}{e^{-x}+1}\, dx$$

$u = e^{-x} + 1$

$e^{-x} = u - 1$

$du = -e^{-x}dx$

$$= -\int \frac{e^{-x} \times -e^{-x}}{e^{-x}+1}\, dx$$

$$= -\int \frac{u-1}{u}\, du$$

$$= -\int \left(1 - \frac{1}{u}\right) du$$

$$= -(u - \ln u) + c$$

$u = e^{-x} + 1 > 0$

$$= \ln u - u + c$$

$$= \ln(e^{-x}+1) - (e^{-x}+1) + c$$

Hint
You can check answers by differentiating.

19 B

Hint
Look for opportunities to complete the square in the denominator to give an inverse trigonometric function.

$$f'(x) = \frac{1}{\sqrt{3-2x-x^2}}$$

$$= \frac{1}{\sqrt{4-(1+2x+x^2)}}$$

$$= \frac{1}{\sqrt{4-(x+1)^2}}$$

$$f(x) = \int \frac{1}{\sqrt{4-(x+1)^2}}\, dx$$

$$= \sin^{-1}\left(\frac{x+1}{2}\right) + c$$

$f(1) = \frac{\pi}{4}$:

$$f(1) = \sin^{-1}\left(\frac{1+1}{2}\right) + c$$

$$\frac{\pi}{4} = \sin^{-1}1 + c$$

$$c = \frac{\pi}{4} - \frac{\pi}{2}$$

$$= -\frac{\pi}{4}$$

So $f(x) = \sin^{-1}\left(\frac{x+1}{2}\right) - \frac{\pi}{4}$.

20 C

$y = \tan^{-1}x$

$x = \tan y$

$$V = \pi\int x^2\, dy$$

$$= \pi\int_0^{\frac{\pi}{3}} \tan^2 y\, dy$$

$$= \pi\int_0^{\frac{\pi}{3}} \sec^2 y - 1\, dy$$

$$= \pi\left[\tan y - y\right]_0^{\frac{\pi}{3}}$$

$$= \pi\left(\tan\frac{\pi}{3} - \frac{\pi}{3} - (0-0)\right)$$

$$= \pi\left(\sqrt{3} - \frac{\pi}{3}\right)$$

$$= \frac{\pi}{3}(3\sqrt{3} - \pi)\ \text{units}^3$$

WORKED SOLUTIONS

Practice set 2

Worked solutions

Question 1

a $y = \sin^{-1} 2x$

$$\frac{dy}{dx} = \frac{1}{\sqrt{1-(2x)^2}} \times 2$$
$$= \frac{2}{\sqrt{1-4x^2}}$$

b $y = \ln(\sin^{-1} 3x)$

$$\frac{dy}{dx} = \frac{1}{\sin^{-1} 3x} \times \frac{1}{\sqrt{1-(3x)^2}} \times 3$$
$$= \frac{3}{\sin^{-1} 3x\sqrt{1-9x^2}}$$

c $y = e^{\cos^{-1} x}$

$$\frac{dy}{dx} = e^{\cos^{-1} x} \times -\frac{1}{\sqrt{1-x^2}}$$
$$= -\frac{e^{\cos^{-1} x}}{\sqrt{1-x^2}}$$

d $y = (\tan^{-1} x)^3$

$$\frac{dy}{dx} = 3(\tan^{-1} x)^2 \times \frac{1}{1+x^2}$$
$$= \frac{3(\tan^{-1} x)^2}{1+x^2}$$

e $y = \tan^{-1}\left(\frac{2}{x}\right)$

$$\frac{dy}{dx} = \frac{1}{1+\left(\frac{2}{x}\right)^2} \times -\frac{2}{x^2}$$
$$= -\frac{2}{x^2\left(1+\frac{4}{x^2}\right)}$$
$$= -\frac{2}{x^2+4}$$

Question 2

a $\int_0^1 \frac{1}{16+x^2}\,dx$

$$= \frac{1}{4}\left[\tan^{-1}\left(\frac{x}{4}\right)\right]_0^1$$
$$= \frac{1}{4}\left(\tan^{-1}\left(\frac{1}{4}\right) - \tan^{-1} 0\right)$$
$$= \frac{1}{4}\tan^{-1}\left(\frac{1}{4}\right)$$

b $\int_{-\sqrt{3}}^0 \frac{1}{\sqrt{4-x^2}}\,dx$

$$= \left[\sin^{-1}\left(\frac{x}{2}\right)\right]_{-\sqrt{3}}^0$$
$$= \sin^{-1} 0 - \sin^{-1}\left(-\frac{\sqrt{3}}{2}\right)$$
$$= 0 - \left(-\frac{\pi}{3}\right)$$
$$= \frac{\pi}{3}$$

Question 3

a $\int \frac{1}{16+9x^2}\,dx$

$$= \frac{1}{4} \times \frac{1}{3}\tan^{-1}\left(\frac{3x}{4}\right) + c$$
$$= \frac{1}{12}\tan^{-1}\left(\frac{3x}{4}\right) + c$$

b $\int \frac{1}{4x^2+4x+2}\,dx$

$$= \int \frac{1}{4x^2+4x+1+1}\,dx$$
$$= \int \frac{1}{(2x+1)^2+1}\,dx$$
$$= \frac{1}{2}\tan^{-1}(2x+1) + c$$

Question 4

a $\text{RHS} = 16 - (2x-3)^2$

$$= 16 - (4x^2 - 12x + 9)$$
$$= 16 - 4x^2 + 12x - 9$$
$$= 7 + 12x - 4x^2$$
$$= \text{LHS}$$

b $\int_0^2 \frac{1}{\sqrt{7+12x-4x^2}}\,dx$

$$= \int_0^2 \frac{1}{\sqrt{16-(2x-3)^2}}\,dx$$
$$= \left[\frac{1}{2}\sin^{-1}\left(\frac{2x-3}{4}\right)\right]_0^2$$
$$= \frac{1}{2}\left[\sin^{-1}\left(\frac{4-3}{4}\right) - \sin^{-1}\left(\frac{0-3}{4}\right)\right]$$
$$= \frac{1}{2}\left[\sin^{-1}\left(\frac{1}{4}\right) - \sin^{-1}\left(-\frac{3}{4}\right)\right]$$

Question 5

$$\int_0^{\frac{\pi}{2}} \cos 5x \cos 2x \, dx$$
$$= \frac{1}{2}\int_0^{\frac{\pi}{2}} \cos(5x - 2x) + \cos(5x + 2x)\, dx$$
$$= \frac{1}{2}\int_0^{\frac{\pi}{2}} \cos 3x + \cos 7x \, dx$$
$$= \frac{1}{2}\left[\frac{1}{3}\sin 3x + \frac{1}{7}\sin 7x\right]_0^{\frac{\pi}{2}}$$
$$= \frac{1}{2}\left(\frac{1}{3}\sin\frac{3\pi}{2} + \frac{1}{7}\sin\frac{7\pi}{2} - (0 + 0)\right)$$
$$= \frac{1}{2}\left(-\frac{1}{3} - \frac{1}{7}\right)$$
$$= -\frac{5}{21}$$

Question 6

$$y = \sin^{-1} 3x$$
$$\frac{dy}{dx} = \frac{1}{\sqrt{1 - (3x)^2}} \times 3$$
$$= \frac{3}{\sqrt{1 - 9x^2}}$$

When $x = \frac{\sqrt{3}}{6}$,

$$\frac{dy}{dx} = \frac{3}{\sqrt{1 - 9 \times \frac{3}{36}}}$$
$$= \frac{3}{\sqrt{\frac{1}{4}}}$$
$$= 6$$

Question 7

$$y = \sqrt{x - 1}$$
$$y^2 = x - 1$$
$$x = y^2 + 1$$
$$V = \int \pi x^2 \, dy$$
$$= \pi\int_0^2 (y^2 + 1)^2 \, dy$$
$$= \pi\int_0^2 y^4 + 2y^2 + 1 \, dy$$
$$= \pi\left[\frac{y^5}{5} + \frac{2y^3}{3} + y\right]_0^2$$
$$= \pi\left[\frac{2^5}{5} + \frac{2 \times 2^3}{3} + 2 - (0 + 0 + 0)\right]$$
$$= \frac{206\pi}{15} \text{ units}^3$$

Question 8

a Using reverse chain rule:

$$\int \sin^3\theta\cos^5\theta \, d\theta$$
$$= \int \sin\theta\sin^2\theta\cos^5\theta \, d\theta$$
$$= \int \sin\theta(1 - \cos^2\theta)\cos^5\theta \, d\theta$$
$$= \int \cos^5\theta\sin\theta - \cos^7\theta\sin\theta \, d\theta$$
$$= -\frac{1}{6}\cos^6\theta + \frac{1}{8}\cos^8\theta + c$$

or using integration by substitution:

$u = \cos\theta$

$du = -\sin\theta \, d\theta$

$-du = \sin\theta \, d\theta$

$$\int \sin^3\theta\cos^5\theta \, d\theta$$
$$= \int \sin\theta\sin^2\theta\cos^5\theta \, d\theta$$
$$= \int (1 - \cos^2\theta)\cos^5\theta\sin\theta \, d\theta$$
$$= \int (\cos^5\theta - \cos^7\theta)\sin\theta \, d\theta$$
$$= \int (u^5 - u^7) \times -du$$
$$= \int (u^7 - u^5)\, du$$
$$= \frac{1}{8}u^8 - \frac{1}{6}u^6 + c$$
$$= \frac{1}{8}\cos^8\theta - \frac{1}{6}\cos^6\theta + c$$

An alternative solution uses $\sin\theta$ as the substitution:

$u = \sin\theta$

$du = \cos\theta \, d\theta$

$$\int \sin^3\theta\cos^5\theta \, d\theta$$
$$= \int \sin^3\theta\cos^4\theta\cos\theta \, d\theta$$
$$= \int \sin^3\theta(1 - \sin^2\theta)^2\cos\theta \, d\theta$$
$$= \int u^3(1 - u^2)^2 \, du$$
$$= \int u^3(1 - 2u^2 + u^4)\, du$$
$$= \int u^3 - 2u^5 + u^7 \, du$$
$$= \frac{1}{4}u^4 - \frac{2}{6}u^6 + \frac{1}{8}u^8 + c$$
$$= \frac{1}{4}\sin^4\theta - \frac{1}{3}\sin^6\theta + \frac{1}{8}\sin^8\theta + c$$

Both solutions are correct although in different forms.

b

$u = \sec\theta$

$du = \sec\theta\tan\theta \, d\theta$

$$\int \tan\theta\sec^3\theta \, d\theta$$
$$= \int \tan\theta\sec\theta\sec^2\theta \, d\theta$$
$$= \int u^2 \, du$$
$$= \frac{1}{3}u^3 + c$$
$$= \frac{1}{3}\sec^3\theta + c$$

WORKED SOLUTIONS

c $\int \cos^4 3\theta \, d\theta$

$= \int (\cos^2 3\theta)^2 \, d\theta$

$= \int \left(\frac{1}{2}(1 + \cos 6\theta)\right)^2 d\theta$

$= \frac{1}{4}\int 1 + 2\cos 6\theta + \cos^2 6\theta \, d\theta$

$= \frac{1}{4}\int 1 + 2\cos 6\theta + \frac{1}{2}(1 + \cos 12\theta) \, d\theta$

$= \frac{1}{4}\int \frac{3}{2} + 2\cos 6\theta + \frac{1}{2}\cos 12\theta \, d\theta$

$= \frac{1}{4}\left(\frac{3x}{2} + \frac{2}{6}\sin 6\theta + \frac{1}{24}\sin 12\theta\right) + c$

$= \frac{3x}{8} + \frac{1}{12}\sin 6\theta + \frac{1}{96}\sin 12\theta + c$

Question 9

a $\int_{\frac{\pi}{4}}^{\frac{\pi}{2}} \cot^2 x \, dx$

$= \int_{\frac{\pi}{4}}^{\frac{\pi}{2}} \operatorname{cosec}^2 x - 1 \, dx$

$= [-\cot x - x]_{\frac{\pi}{4}}^{\frac{\pi}{2}}$

$= -\cot\frac{\pi}{2} - \frac{\pi}{2} - \left(-\cot\frac{\pi}{4} - \frac{\pi}{4}\right)$

$= 0 - \frac{\pi}{2} + \sqrt{2} + \frac{\pi}{4}$

$= \sqrt{2} - \frac{\pi}{4}$

b $\int_0^{\frac{\pi}{4}} \tan^2\theta \sec^4\theta \, d\theta$

$= \int_0^{\frac{\pi}{4}} \tan^2\theta \sec^2\theta \sec^2\theta \, d\theta$

$= \int_0^{\frac{\pi}{4}} \tan^2\theta(1 + \tan^2\theta)\sec^2\theta \, d\theta$

$= \int_0^1 u^2 + u^4 \, du$ $\quad u = \tan\theta$

$= \left[\frac{1}{3}u^3 + \frac{1}{5}u^5\right]_0^1$ $\quad du = \sec^2\theta$, $\theta = 0, u = 0$

$= \left(\frac{1}{3} + \frac{1}{5} - (0 + 0)\right)$ $\quad \theta = \frac{\pi}{4}, u = 1$

$= \frac{8}{15}$

c $\int_{\frac{\pi}{2}}^{2\pi} 4\sin^2\left(\frac{x}{4}\right)\cos^2\left(\frac{x}{4}\right) dx$

$= \int_{\frac{\pi}{2}}^{2\pi} \left(2\sin\frac{x}{4}\cos\frac{x}{4}\right)^2 dx$

$= \int_{\frac{\pi}{2}}^{2\pi} \sin^2\frac{x}{2} \, dx$

$= \frac{1}{2}\int_{\frac{\pi}{2}}^{2\pi} 1 - \cos x \, dx$

$= \frac{1}{2}[x - \sin x]_{\frac{\pi}{2}}^{2\pi}$

$= \frac{1}{2}\left(2\pi - \sin 2\pi - \left(\frac{\pi}{2} - \sin\frac{\pi}{2}\right)\right)$

$= \frac{1}{2}\left(2\pi - \frac{\pi}{2} + 1\right)$

$= \frac{3\pi}{4} + \frac{1}{2}$

Question 10

a LHS $= 4\sin 2x \sin 4x \sin 6x$

$= 2\sin 6x \sin 2x(2\sin 4x)$

$= [\cos(6x - 2x) - \cos(6x + 2x)](2\sin 4x)$

$= [\cos 4x - \cos 8x](2\sin 4x)$

$= 2\sin 4x\cos 4x - 2\sin 4x\cos 8x$

$= \sin 8x - [\sin(8x + 4x) - \sin(8x - 4x)]$

$= \sin 8x - \sin 12x + \sin 4x$

$=$ RHS

b $\int \sin 2x \sin 4x \sin 6x \, dx$

$= \frac{1}{4}\int 4\sin 2x \sin 4x \sin 6x \, dx$

$= \frac{1}{4}\int \sin 4x + \sin 8x - \sin 12x \, dx$

$= \frac{1}{4}\left(-\frac{1}{4}\cos 4x - \frac{1}{8}\cos 8x + \frac{1}{12}\cos 12x\right) + c$

$= \frac{1}{48}\cos 12x - \frac{1}{32}\cos 8x - \frac{1}{16}\cos 4x + c$

WORKED SOLUTIONS

Question 11

a $\text{RHS} = 2 - \dfrac{2x+7}{x^2+1}$

$= \dfrac{2(x^2+1)-(2x+7)}{x^2+1}$

$= \dfrac{2x^2+2-2x-7}{x^2+1}$

$= \dfrac{2x^2-2x-5}{x^2+1}$

$= \text{LHS}$

b $\int_1^6 \dfrac{2x^2-2x-5}{x^2+1}\,dx$

$= \int_1^6 2 - \dfrac{2x+7}{x^2+1}\,dx$

$= \int_1^6 2 - \dfrac{2x}{x^2+1} - \dfrac{7}{x^2+1}\,dx$

$= \left[2x - \ln(x^2+1) - 7\tan^{-1}x\right]_1^6 \quad x^2+1>0$

$= (12 - \ln 37 - 7\tan^{-1}6) - (2 - \ln 2 - 7\tan^{-1}1)$

$= 12 - \ln 37 - 7\tan^{-1}6 - 2 + \ln 2 + \dfrac{7\pi}{4}$

≈ 2.74

Question 12

a $\int \dfrac{\tan^{-1}x}{1+x^2}\,dx \qquad u = \tan^{-1}x, \quad du = \dfrac{dx}{1+x^2}$

$= \int u\,du$

$= \dfrac{1}{2}u^2 + c$

$= \dfrac{1}{2}(\tan^{-1}x)^2 + c$

b $\int \dfrac{x+1}{\sqrt{x^2+2x}}\,dx \qquad u = x^2+2x, \quad du = (2x+2)\,dx, \quad \dfrac{1}{2}du = (x+1)\,dx$

$= \int \dfrac{1}{\sqrt{u}} \times \dfrac{1}{2}\,du$

$= \dfrac{1}{2}\int u^{-\frac{1}{2}}\,du$

$= \dfrac{1}{2} \times \dfrac{u^{\frac{1}{2}}}{\frac{1}{2}} + c$

$= u^{\frac{1}{2}} + c$

$= \sqrt{x^2+2x} + c$

c $\int_0^1 x\sqrt{1-x}\,dx \qquad u = 1-x, \ x = 1-u, \quad du = -dx, \quad x=0,\ u=1, \quad x=1,\ u=0$

$= \int_1^0 (1-u)\sqrt{u} \times -du$

$= \int_0^1 u^{\frac{1}{2}} - u^{\frac{3}{2}}\,du$

$= \left[\dfrac{2}{3}u^{\frac{3}{2}} - \dfrac{2}{5}u^{\frac{5}{2}}\right]_0^1$

$= \dfrac{2}{3} \times 1^{\frac{3}{2}} - \dfrac{2}{5} \times 1^{\frac{5}{2}} - (0-0)$

$= \dfrac{2}{3} - \dfrac{2}{5}$

$= \dfrac{4}{15}$

Question 13

$\dfrac{d^2y}{dx^2} = x(9-x^2)^{-\frac{3}{2}}$

$\dfrac{dy}{dx} = \int x(9-x^2)^{-\frac{3}{2}}\,dx$

$= -\dfrac{1}{2}\int -2x(9-x^2)^{-\frac{3}{2}}\,dx$

$= -\dfrac{1}{2} \times \dfrac{(9-x^2)^{-\frac{1}{2}}}{-\frac{1}{2}} + c_1$

$= \dfrac{1}{\sqrt{9-x^2}} + c_1$

When $x = 0$, $\dfrac{dy}{dx} = \dfrac{1}{3}$:

$\dfrac{1}{3} = \dfrac{1}{\sqrt{9-0}} + c_1$

$c_1 = 0$:

So $\dfrac{dy}{dx} = \dfrac{1}{\sqrt{9-x^2}}$

$y = \int \dfrac{1}{\sqrt{9-x^2}}\,dx$

$= \sin^{-1}\left(\dfrac{x}{3}\right) + c_2$

When $x = 0$, $y = \pi$:

$\pi = \sin^{-1}0 + c_2$

$c_2 = \pi$

So $y = \sin^{-1}\left(\dfrac{x}{3}\right) + \pi$.

Question 14

a $\frac{d}{dx}\sin(\cos^{-1} x)$

$= \cos(\cos^{-1} x) \times -\frac{1}{\sqrt{1-x^2}}$

$= -\frac{x}{\sqrt{1-x^2}}$

b If $\frac{d}{dx}\sin(\cos^{-1} x) = -\frac{x}{\sqrt{1-x^2}}$,

then $\int -\frac{x}{\sqrt{1-x^2}}\, dx = \sin(\cos^{-1} x)$.

$\text{LHS} = \frac{1}{2}\int -2x(1-x^2)^{-\frac{1}{2}}\, dx$

$= \frac{1}{2} \times \frac{(1-x^2)^{\frac{1}{2}}}{\frac{1}{2}} + c$

$= \sqrt{1-x^2} + c$

but if $y = \sin(\cos^{-1} x)$, $x = 0$ and $y = 1$.

Substitute this condition into the new function:

$y = \sqrt{1-x^2} + c$

$1 = \sqrt{1-0} + c$

$c = 0$

So $y = \sqrt{1-x^2}$
$= \text{RHS}$

$\sin(\cos^{-1} x) = \sqrt{1-x^2}$

Alternative solution (without using part **a**):

Let $\cos^{-1} x = \theta$,

then $x = \cos\theta$, where $0 \le \theta \le \pi$.

$\sin\theta = \pm\sqrt{1-\cos^2\theta}$ but since $0 \le \theta \le \pi$,

$= \sqrt{1-\cos^2\theta}$ $\sin\theta \ge 0$

$= \sqrt{1-x^2}$

$\therefore \sin(\cos^{-1} x) = \sqrt{1-x^2}$

Question 15

a Find points of intersection by solving simultaneously:

$x + y = 5$ becomes $y = 5 - x$ [1]

$xy = 4$ [2]

Substitute [1] into [2]:

$x(5-x) = 4$

$5x - x^2 = 4$

$x^2 - 5x + 4 = 0$

$(x-1)(x-4) = 0$

$\therefore x = 1$ or $x = 4$

When $x = 1$, $y = 4$.

When $x = 4$, $y = 1$.

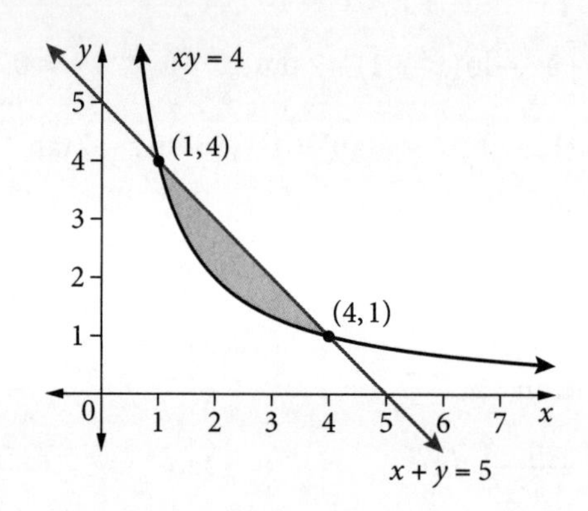

b $A = \int_1^4 5 - x - \frac{4}{x}\, dx$

$= \left[5x - \frac{1}{2}x^2 - 4\ln|x|\right]_1^4$

$= \left(20 - \frac{1}{2}\times 16 - 4\ln 4\right) - \left(5 - \frac{1}{2}\times 1 - 4\ln 1\right)$

$= 20 - 8 - 4\ln 4 - 5 + \frac{1}{2} + 0$

$= \frac{15}{2} - 4\ln 4$

$= \frac{15}{2} - 8\ln 2 \text{ units}^2$

c $V = \pi\int_1^4 (5-x)^2\,dx - \pi\int_1^4 \left(\frac{4}{x}\right)^2 dx$

$= \pi\int_1^4 (25 - 10x + x^2)\,dx - \pi\int_1^4 \frac{16}{x^2}\,dx$

$= \pi\left[25x - 5x^2 + \frac{1}{3}x^3\right]_1^4 - \pi\left[-\frac{16}{x}\right]_1^4$

$= \pi\left[\left(25\times 4 - 5\times 4^2 + \frac{1}{3}\times 4^3\right) - \left(25 - 5 + \frac{1}{3}\right)\right] + \pi\left(\frac{16}{4} - \frac{16}{1}\right)$

$= \pi\left(\frac{124}{3} - \frac{61}{3}\right) - 12\pi$

$= 9\pi$ units^3

Question 16

a $y = x\sin^{-1}x$

$\frac{dy}{dx} = x \times \frac{1}{\sqrt{1-x^2}} + \sin^{-1}x$

$= \frac{x}{\sqrt{1-x^2}} + \sin^{-1}x$

b If $\frac{d}{dx}x\sin^{-1}x = \frac{x}{\sqrt{1-x^2}} + \sin^{-1}x$,

then $\int \frac{x}{\sqrt{1-x^2}}\,dx + \int \sin^{-1}x\,dx = x\sin^{-1}x + c$

$\int \sin^{-1}x\,dx = x\sin^{-1}x - \int \frac{x}{\sqrt{1-x^2}}\,dx$

$= x\sin^{-1}x + \frac{1}{2}\int \frac{-2x}{\sqrt{1-x^2}}\,dx$

$= x\sin^{-1}x + \sqrt{1-x^2} + c$

Question 17

a $\int (1-x^2)^{\frac{3}{2}}\,dx$ $\quad x = \sin\theta$, $dx = \cos\theta\,d\theta$

$= \int (1-\sin^2\theta)^{\frac{3}{2}} \times \cos\theta\,d\theta$

$= \int (\cos^2\theta)^{\frac{3}{2}} \times \cos\theta\,d\theta$

$= \int \cos^4\theta\,d\theta$

$= \int \left(\frac{1}{2}(1+\cos 2\theta)\right)^2 d\theta$

$= \frac{1}{4}\int 1 + 2\cos 2\theta + \cos^2 2\theta\,d\theta$

$= \frac{1}{4}\int 1 + 2\cos 2\theta + \frac{1}{2} + \frac{1}{2}\cos 4\theta\,d\theta$

$= \frac{1}{4}\int \frac{3}{2} + 2\cos 2\theta + \frac{1}{2}\cos 4\theta\,d\theta$

$= \frac{1}{4}\left[\frac{3\theta}{2} + \sin 2\theta + \frac{1}{8}\sin 4\theta\right] + c$

$= \frac{1}{32}(12\theta + 8\sin 2\theta + \sin 4\theta) + c$

b $\int \frac{1}{(16+x^2)^{\frac{3}{2}}}\,dx$ $\quad x = 4\tan\theta$, $dx = 4\sec^2\theta\,d\theta$

$= \int \frac{1}{(16 + 16\tan^2\theta)^{\frac{3}{2}}} \times \sec^2\theta\,d\theta$

$= \int \frac{\sec^2\theta}{(16(1+\tan^2\theta))^{\frac{3}{2}}}\,d\theta$

$= \int \frac{\sec^2\theta}{64\sec^3\theta}\,d\theta$

$= \frac{1}{64}\int \frac{1}{\sec\theta}\,d\theta$

$= \frac{1}{64}\int \cos\theta\,d\theta$

$= \frac{1}{64}\sin\theta + c$

$= \frac{1}{64} \times \frac{x}{\sqrt{16+x^2}} + c$

$= \frac{x}{64\sqrt{16+x^2}} + c$

c $\int_{-1}^{0} \frac{x+1}{\sqrt{1-x}}\,dx$

$u = \sqrt{1-x}$

Also, $u^2 = 1 - x$

$x = 1 - u^2$

$dx = -2u\,du$

$x = -1, u = \sqrt{2}$

$x = 0, u = 1$

$= \int_{\sqrt{2}}^{1} \frac{1-u^2+1}{u}(-2u)\,du$

$= -2\int_{\sqrt{2}}^{1} 2 - u^2\,du$

$= 2\int_{1}^{\sqrt{2}} 2 - u^2\,du$

$= 2\left[2u - \frac{1}{3}u^3\right]_1^{\sqrt{2}}$

$= 2\left(2\sqrt{2} - \frac{1}{3}\times\sqrt{2}^3 - \left(2 - \frac{1}{3}\right)\right)$

$= 2\left(\frac{4\sqrt{2}}{3} - \frac{5}{3}\right)$

$= \frac{2}{3}(4\sqrt{2} - 5)$

Question 18

a $\int x(1-x^2)^4\,dx$

Let $u = 1 - x^2$

$du = -2x\,dx$

$-\frac{1}{2x}du = x\,dx$

$= \int xu^4 \times -\frac{1}{2x}\,du$

$= -\frac{1}{2}\times\frac{1}{5}u^5 + c$

$= -\frac{1}{10}u^5 + c$

$= -\frac{1}{10}(1-x^2)^5 + c$

b $\int \frac{1}{x\ln x}\,dx$

Let $u = \ln x$

$du = \frac{1}{x}dx$

$= \int \frac{1}{\ln x}\times\frac{1}{x}\,dx$

$= \int \frac{1}{u}\,du$

$= \ln|u| + c$

$= \ln|\ln x| + c$

c $\int \frac{x}{4+9x^4}\,dx$

Let $u = 3x^2$

$du = 6x\,dx$

$\frac{1}{6}du = x\,dx$

$= \frac{1}{6}\int \frac{1}{4+u^2}\,du$

$= \frac{1}{6}\times\frac{1}{2}\tan^{-1}\left(\frac{u}{2}\right) + c$

$= \frac{1}{12}\tan^{-1}\left(\frac{u}{2}\right) + c$

$= \frac{1}{12}\tan^{-1}\left(\frac{3x^2}{2}\right) + c$

Question 19

a Use factor theorem to determine a zero:

$f(2) = 2^3 - 2^2 - 8(2) + 12 = 0$

So $(x-2)$ is a zero of $f(x)$.

$$
\begin{array}{r}
x^2 + x - 6 \\
x - 2\,\overline{)\,x^3 - x^2 - 8x + 12} \\
\underline{x^3 - 2x^2} \\
x^2 - 8x + 12 \\
\underline{x^2 - 2x} \\
-6x + 12 \\
\underline{-6x + 12} \\
0
\end{array}
$$

$f(x) = (x-2)(x^2+x-6)$

$= (x-2)(x+3)(x-2)$

$= (x-2)^2(x+3)$

Note: Skills learnt in Year 11 Further graphing are required to complete the graph of the square root function.

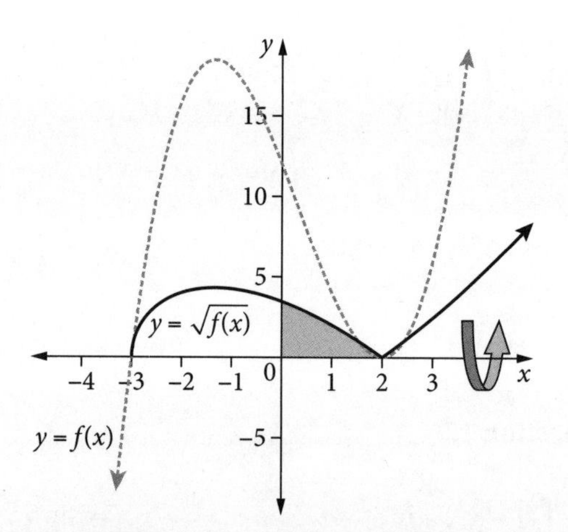

b $V = \pi\int_0^2 \left(\sqrt{x^3 - x^2 - 8x + 12}\right)^2 dx$

$= \pi\int_0^2 x^3 - x^2 - 8x + 12\,dx$

$= \pi\left[\frac{1}{4}x^4 - \frac{1}{3}x^3 - 4x^2 + 12x\right]_0^2$

$= \pi\left(\frac{2^4}{4} - \frac{2^3}{3} - 4(2^2) + 12(2) - (0 - 0 - 0 + 0)\right)$

$= \frac{28\pi}{3}$ units^3

Question 20

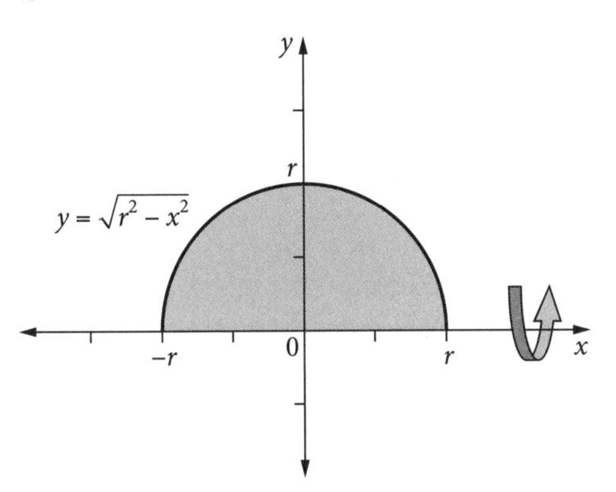

$$V = \int \pi y^2\, dx$$
$$= \pi \int_{-r}^{r} \left(\sqrt{r^2 - x^2}\right)^2 dx$$
$$= \pi \int_{-r}^{r} r^2 - x^2\, dx$$
$$= \pi \left[r^2 x - \frac{1}{3}x^3 \right]_{-r}^{r}$$
$$= \pi \left(r^3 - \frac{1}{3}r^3 - \left(-r^3 + \frac{1}{3}r^3\right)\right)$$
$$= \pi \left(\frac{2}{3}r^3 - \left(-\frac{2}{3}r^3\right)\right)$$
$$= \frac{4}{3}\pi r^3 \text{ units}^3$$

Question 21

a

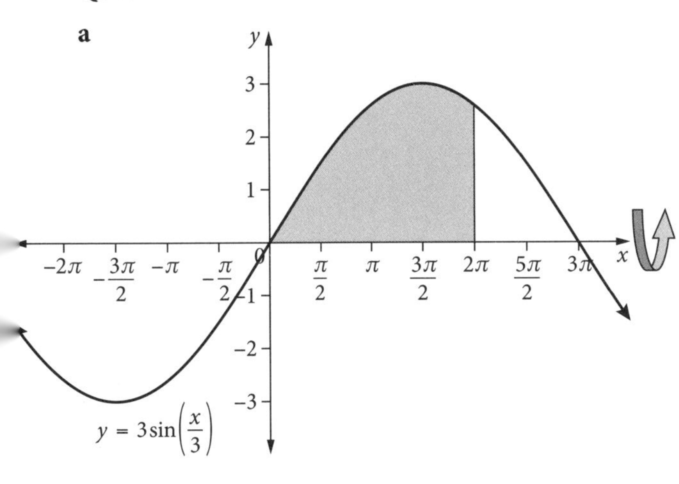

$$V = \pi \int_0^{2\pi} \left(3\sin\frac{x}{3}\right)^2 dx$$
$$= \pi \int_0^{2\pi} 9\sin^2\frac{x}{3}\, dx$$
$$= \frac{9\pi}{2} \int_0^{2\pi} 1 - \cos\frac{2x}{3}\, dx$$
$$= \frac{9\pi}{2} \left[x - \frac{3}{2}\sin\frac{2x}{3} \right]_0^{2\pi}$$
$$= \frac{9\pi}{2} \left(2\pi - \frac{3}{2}\sin\frac{4\pi}{3} - (0 - 0)\right)$$
$$= \frac{9\pi}{2} \left(2\pi + \frac{3\sqrt{3}}{4}\right)$$
$$= \frac{9\pi}{8}(8\pi + 3\sqrt{3}) \text{ units}^3$$

b $V = \pi \int_0^{h} \left(3\sin\frac{x}{3}\right)^2 dx$

$$= \frac{9\pi}{2} \left[x - \frac{3}{2}\sin\left(\frac{2x}{3}\right) \right]_0^{h} \qquad \text{(from part } \mathbf{a}\text{)}$$
$$= \frac{9\pi}{2} \left(h - \frac{3}{2}\sin\left(\frac{2h}{3}\right) - (0 - 0)\right)$$
$$= \frac{9\pi}{4} \left(2h + 3\sin\left(\frac{2h}{3}\right)\right) \text{ units}^3$$

c Find the volume of liquid.

Substitute $h = \pi$ into above formula:

$$V_l = \frac{9\pi}{4} \left(2\pi + 3\sin\left(\frac{2\pi}{3}\right)\right)$$
$$= \frac{9\pi}{4} \left(2\pi + \frac{3\sqrt{3}}{2}\right)$$
$$= \frac{9\pi}{8}(4\pi + 3\sqrt{3}) \text{ units}^3$$

Find the volume of air (use part **a**).

$$V_a = \frac{9\pi}{8}(8\pi + 3\sqrt{3}) - \frac{9\pi}{8}(4\pi + 3\sqrt{3})$$

or integrate from $x = 2\pi$ to 3π.

$$= \frac{9\pi}{8}(8\pi + 3\sqrt{3} - 4\pi - 3\sqrt{3})$$
$$= \frac{9\pi}{8}(4\pi) \text{ units}^3$$

Ratio of liquid to air:

$$V_l : V_a$$
$$\frac{9\pi}{8}(4\pi + 3\sqrt{3}) : \frac{9\pi}{8}(4\pi)$$
$$4\pi + 3\sqrt{3} : 4\pi$$

9780170459242

WORKED SOLUTIONS

HSC exam topic grid (2011–2020)

This grid shows the coverage of this topic in past HSC exams by question number. The past exams can be downloaded from the NESA website (www.educationstandards.nsw.edu.au) by selecting 'Year 11 – Year 12', 'HSC exam papers'. NESA marking feedback and guidelines can also be found there.

The new Mathematics Extension 1 course was first examined in 2020. Volumes of solids of revolution was in the 'Mathematics' course before 2020. For exams before 2020, select 'Year 11 – Year 12', 'Resources archive', 'HSC exam papers archive'.

	Integration by substitution	Trigonometric integrals	Inverse functions and inverse trigonometric functions	Volumes of solids of revolution * Mathematics exam before 2020
2011	1(d)			8(b)*
2012	11(d)	7	9, 11(a)	14(b)*
2013	5, 11(f)	12(b)	11(b), 11(g)	15(b)*, Ext 1: 12(b)
2014	11(d)	12(b)	6	14(c)*, Ext 1: 12(b)
2015	11(e)	11(a)	7, 13(d)	16(b)*
2016	11(b)	5	11(c)	15(a)*
2017	11(e)	11(f)	11(b)	12(b)*, Ext 1:12(c)(i)
2018	11(f)	12(a)	12(c)	14(b)*
2019	13(a)	11(e), 13(a)	3	13(d)*
2020 new course	13(a)	12(d), 13(b)	3, 13(c)	13(b)

9780170459242

CHAPTER 5 DIFFERENTIAL EQUATIONS

ME-C3 Applications of calculus 110

C3.2 Differential equations 110

9780170459242

DIFFERENTIAL EQUATIONS

Solving differential equations

$\frac{dy}{dx} = f(x)$

$\frac{dy}{dx} = g(y)$

$\frac{dy}{dx} = f(x)\,g(y)$: separation of variables

Direction fields

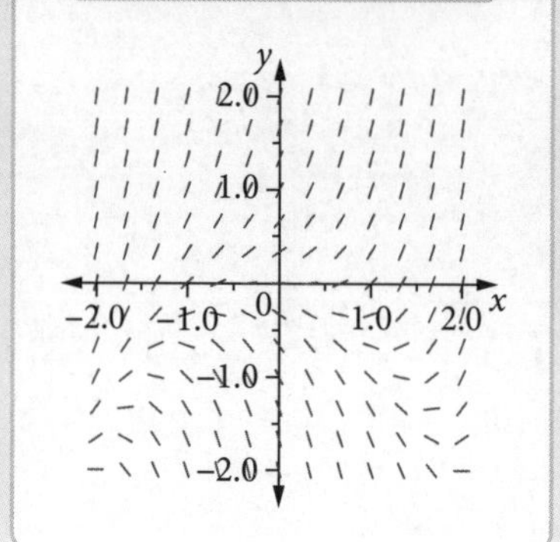

Application of differential equations

- Exponential growth and decay
- Newton's law of cooling
- Logistic equations

Glossary

differential equation
An equation that involves a function and its derivative(s), whose solution is the function.

direction field or slope field
A diagram that shows the gradients of the infinite possible curves on a number plane for a given differential equation.

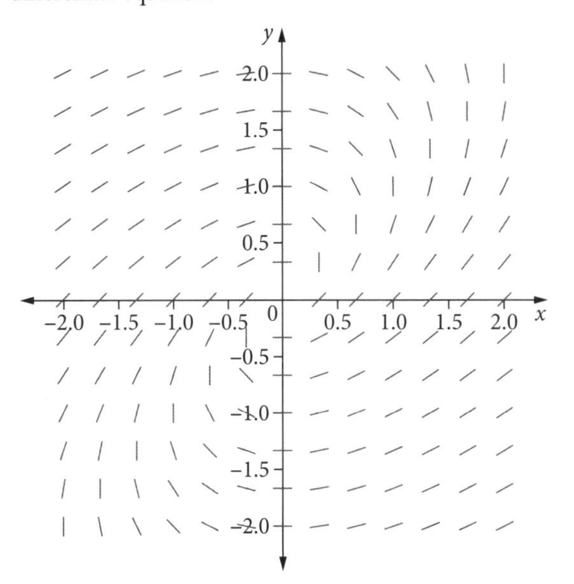

exponential growth and decay
A model of differential equations in which the rate of change of the quantity is proportional to the amount of quantity.

$$\frac{dP}{dt} = kP$$

The solution is an exponential function, $P = Ae^{kt}$.

A+ DIGITAL FLASHCARDS
Revise this topic's key terms and concepts by scanning the QR code or typing the URL into your browser.

https://get.ga/a-hsc-maths-ext-1

logistic equation
A differential equation in which the rate of change of the quantity is proportional to the amount of quantity and the difference between the limiting condition N and the quantity:

$$\frac{dP}{dt} = kP(N - P)$$

Newton's law of cooling
A law that can be described by a differential equation that states that the rate of decrease in a quantity such as temperature is proportional to the difference between the quantity such as room temperature and some constant quantity,

$$\frac{dT}{dt} = k(T - T_1),$$

where $k < 0$.
A typical example is in cooling a liquid from boiling point.

separation of variables
To separate a differential equation written in the form

$$\frac{dy}{dx} = f(x)\,g(y),$$

to the form

$$\frac{1}{g(y)}\,dy = f(x)\,dx$$

for integration.

Topic summary

Applications of calculus (ME-C3)

C3.2 Differential equations

Type 1: $\frac{dy}{dx} = f(x)$

These are the standard **differential equations** that you have been solving since you first learnt to integrate functions.

$$\text{If } \frac{dy}{dx} = f(x), \text{ then } y = \int f(x)\,dx.$$

Type 2: $\frac{dy}{dx} = g(y)$

When the derivative is given as a function of y, take the reciprocal of both sides of the differential equation.

We integrate $\frac{dx}{dy}$ with respect to y to get an equation for x as a function of y.

$$\text{If } \frac{dy}{dx} = g(y), \text{ then } \frac{dx}{dy} = \frac{1}{g(y)} \text{ and } x = \int \frac{1}{g(y)}\,dy.$$

Testing for $g(y) = 0$

Because we are dividing both sides by $g(y)$ when solving this differential equation, we must also test separately whether $g(y) = 0$ is a solution.

For example, for the differential equation $\frac{dy}{dx} = y - 4$, we rewrite as $\frac{dx}{dy} = \frac{1}{y-4}$ assuming $y - 4 \neq 0$, so we must test whether $y - 4 = 0$ is a solution:

$$y - 4 = 0$$
$$y = 4$$

$$\begin{aligned}\text{LHS} &= \frac{dy}{dx} \\ &= 0\end{aligned}$$

$$\begin{aligned}\text{RHS} &= y - 4 \\ &= 4 - 4 \\ &= 0 \\ &= \text{LHS}\end{aligned}$$

$\therefore$ $y = 4$ is a solution to this differential equation, and is called the **constant solution** because it is a constant function. On a direction field, the constant solution is represented by flat (horizontal) dashes.

Type 3: $\frac{dy}{dx} = f(x)\,g(y)$

When the derivative is given as a **product** of a function of x and a function of y:

1. **Separate the variables**, including the derivative fraction.
2. Integrate both sides with respect to the part of the derivative.

If $\frac{dy}{dx} = f(x)g(y)$, then we can rewrite as $\frac{1}{g(y)}\,dy = f(x)\,dx$ and integrate both sides.

$$\int \frac{1}{g(y)}\,dy = \int f(x)\,dx$$

Also test whether $g(y) = 0$ is a solution here.

Hint

Even though you are integrating both sides of the equation, only one constant is required (by convention, it is placed on the side with x). This is because if a constant is placed on both sides of an equation, it can be easily removed by subtracting one constant from both sides of the equation.

In this course, we only examine and solve **first-order** differential equations, which are equations involving the **first derivative** only. (Differential equations involving the *second derivative* are called *second-order* differential equations.)

Direction fields

A **direction field** or **slope field** is a graph of possible solutions to a differential equation, represented by dashes of lines or arrows showing the tangents of this 'family of curves'. Direction fields are usually created by technology such as GeoGebra and Desmos. The direction field below also shows 3 possible curves as solutions to the differential equation, created by 'joining the dashes'.

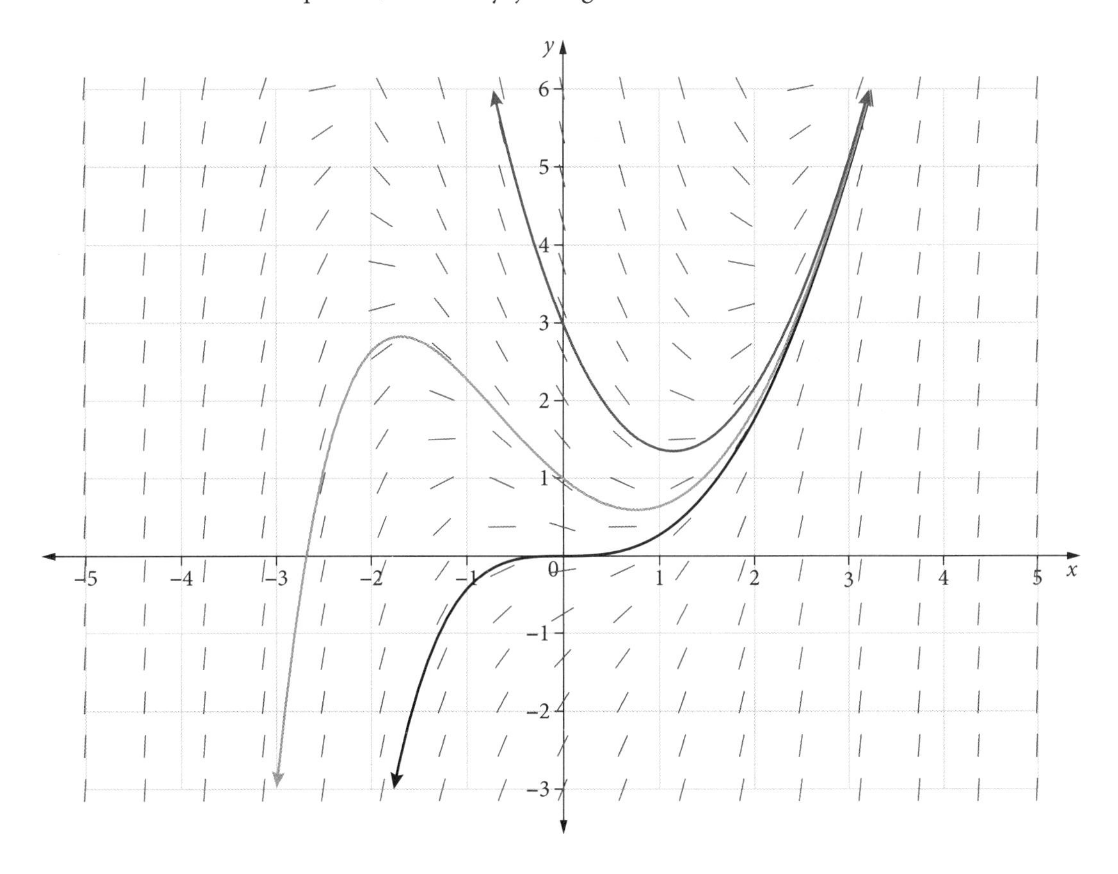

Drawing a direction field in an exam is very time-consuming, so instead you could be asked to draw a possible solution (or several possible solutions) from a given direction field and indicate your solution by identifying key points.

Because you will be writing your answers in a separate exam booklet, it is highly unlikely you will reproduce a complete direction field. Rather, if asked to sketch a particular solution, you may wish to sketch some gradient lines to determine the graph of the solution to the differential equation. A direction field question will most likely be multiple choice, in which you select the correct direction field for a differential equation or the correct differential equation for a direction field. See Question 6 onwards in Practice set 1.

Exponential growth and decay

- Differential equations are of the form $\frac{dP}{dt} = kP$, where k is a constant.
- The solution has the form $P = Ae^{kt}$, where A is the initial value.
  - If $k > 0$, k is the growth factor in exponential growth.
  - If $k < 0$, k is the rate of decay in exponential decay.

Newton's law of cooling

- Differential equations are of the form $\frac{dT}{dt} = k(T - T_1)$, where k is a negative constant ($k < 0$, rate of cooling) and T_1 is the constant quantity (room temperature, limiting condition).
- The solution has the form $T = T_1 + Ae^{kt}$.
- Sometimes, these formulas are written with '$-k$' instead of 'k', where $k > 0$.
- Hot objects lose heat according to this law, where T is the temperature and T_1 is a limiting temperature such as room temperature.

Logistic equation

This differential equation has the form $\frac{dP}{dt} = kP(N - P)$, where k and N are constants.

It is a more realistic model of population growth that does not increase indefinitely, where N is the limiting condition or carrying capacity.

Hint

In solving this differential equation, it is not possible to integrate $\frac{1}{kP(N - P)}$ easily (unless you have learnt partial fractions in Maths Extension 2). In most instances, you will be provided with a method to split this into the sum or difference of 2 fractions, which you can then use to integrate.

TRACKING GRID

Practice sets tracking grid

Maths is all about repetition, meaning do, do and do again! Each question in the following practice sets, especially the struggle questions (different for everybody!), should be completed at least 3 times correctly. Below is a tracking grid to record your question attempts: ✓ if you answered correctly, ✗ if you didn't.

PRACTICE SET 1: Multiple-choice questions

Question	1st attempt	2nd attempt	3rd attempt	4th attempt	5th attempt
1					
2					
3					
4					
5					
6					
7					
8					
9					
10					
11					
12					
13					
14					
15					
16					
17					
18					
19					
20					

PRACTICE SET 2: Short-answer questions

Question	1st attempt	2nd attempt	3rd attempt	4th attempt	5th attempt
1					
2					
3					
4					
5					
6					
7					
8					
9					
10					
11					
12					
13					
14					
15					
16					
17					
18					
19					
20					

Practice set 1

Multiple-choice questions

Solutions start on page 125.

Question 1

Which of the following is NOT a separable differential equation?

A $\dfrac{dy}{dx} = x + y^2$ **B** $\dfrac{dy}{dx} + xy = 4x$ **C** $\dfrac{dy}{dx} = \sqrt{1 + x^2}$ **D** $\dfrac{dy}{dx} = 2^{y-x}$

Question 2

A curve has a differential equation $\dfrac{dy}{dx} = \dfrac{1-x}{2+y}$. The gradient at point P is $\dfrac{2}{9}$.

What are the coordinates of P?

A $(2,7)$ **B** $(-3,-11)$ **C** $(-1,7)$ **D** $(2,15)$

Question 3

Which of the following is a possible solution for the differential equation $\dfrac{dy}{dx} + xy - x^3 = 0$?

A $y = x^2 - 2x$ **B** $y = x^2 - 2$ **C** $y = x^2 + 2x$ **D** $y = x^2 + 2$

Question 4

For which of the following differential equations is $y = e^{-2x}$ a solution?

A $y'' + 2y' = 0$ **B** $y'' + 2y' + 2y = 0$ **C** $y' - 2y = 0$ **D** $2y'' + 3y' + 2y = 0$

Question 5

A curve has a differential equation $\dfrac{dy}{dx} = 3x^2 - 7xy + 2y^2$.

What is the gradient of the curve at the point that passes through $(2,5)$?

A -132 **B** -8 **C** 13 **D** 153

Question 6

Which differential equation gives the direction field shown below?

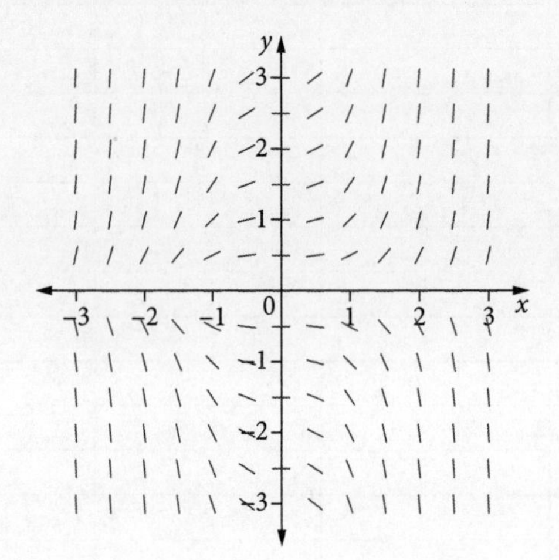

A $\dfrac{dy}{dx} = xy^2$ **B** $\dfrac{dy}{dx} = x^2y$ **C** $\dfrac{dy}{dx} = -xy^2$ **D** $\dfrac{dy}{dx} = -x^2y$

Question 7

The direction field for the differential equation $\frac{dy}{dx} = x^2 + 2y - 2$ is shown below.

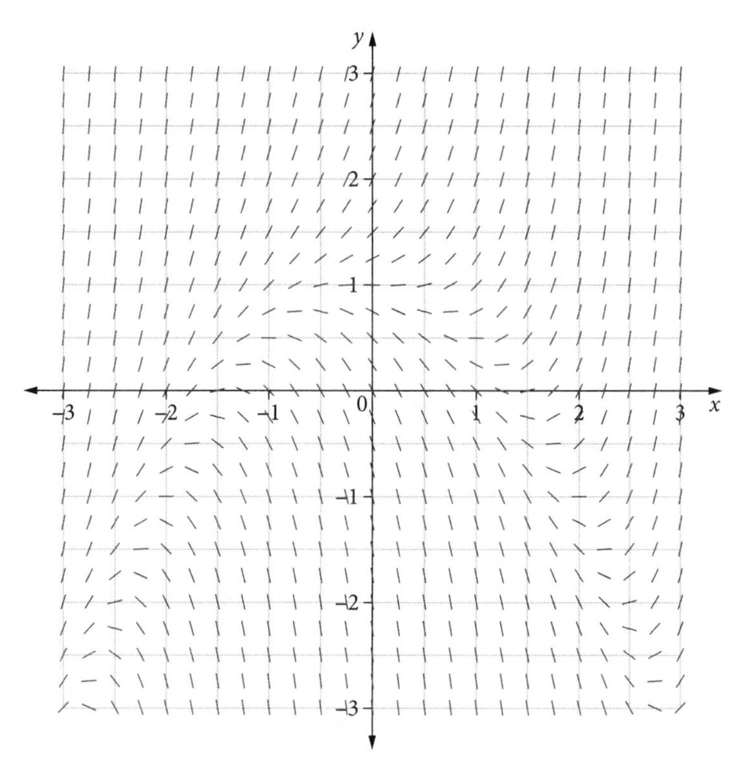

Which of the following points could a solution curve pass through if it also passes through $(1, 0)$?

A $(0.5, -1.5)$

B $(-1.5, 0.5)$

C $(1.5, 1.5)$

D $(-1.5, -0.5)$

Question 8

Which of the following is a possible general solution to the direction field below?

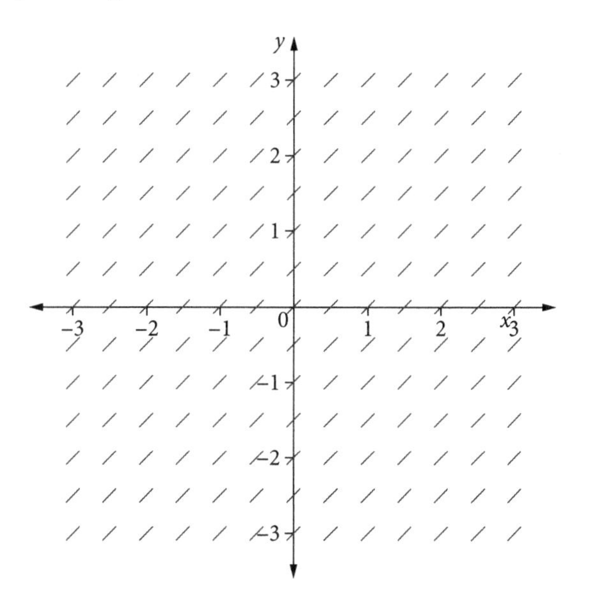

A $y = x + c$

B $y = 0$

C $y = -x + c$

D $x = 0$

Question 9

Which of the following equations is a possible general solution to the direction field below?

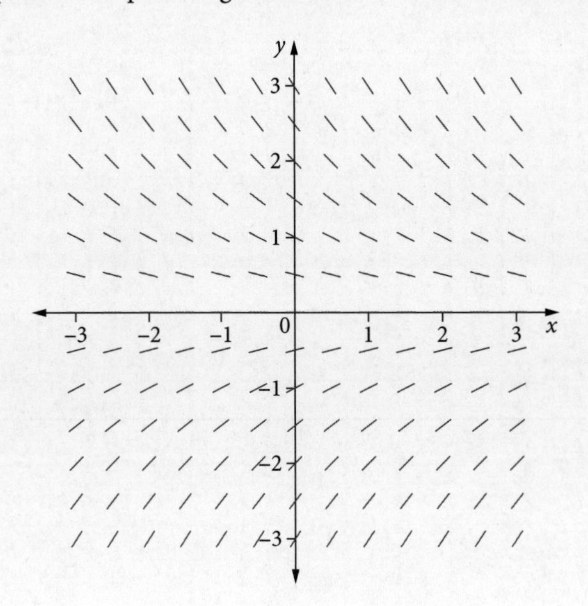

A $y = Ae^x + c$

B $y = -Ae^x + c$

C $y = -Ae^{-x} + c$

D $y = 1$

Question 10

Which differential equation gives the direction field shown below?

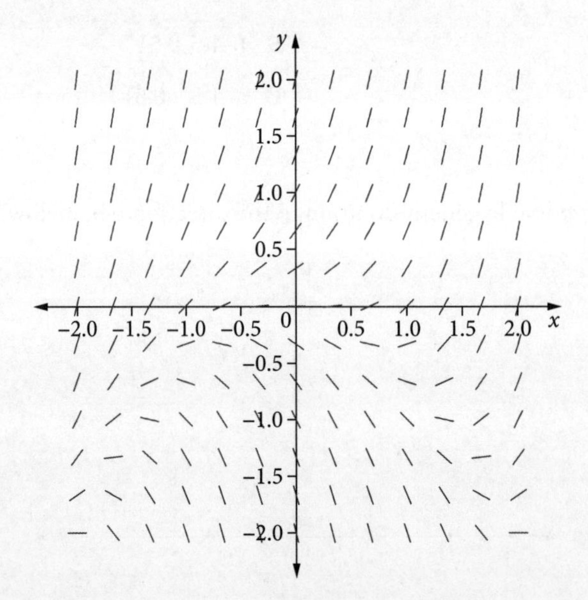

A $\frac{dy}{dx} = x^2 + 2y$

B $\frac{dy}{dx} = 2x + y^2$

C $\frac{dy}{dx} = x^2 - 2y$

D $\frac{dy}{dx} = 2x - y^2$

Question 11

Which direction field best matches $\frac{dy}{dx} = \frac{xy}{2}$?

A

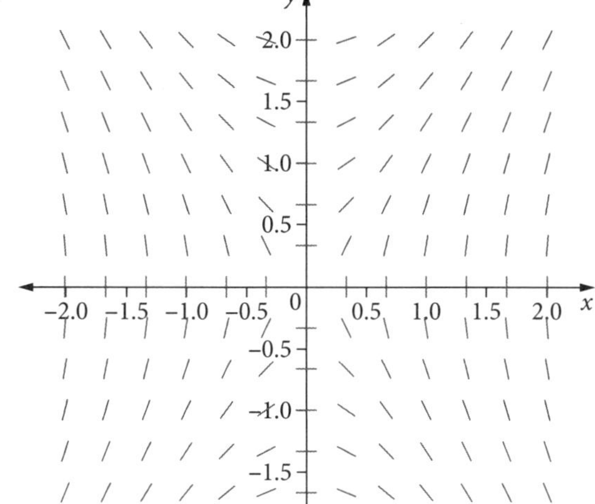

B

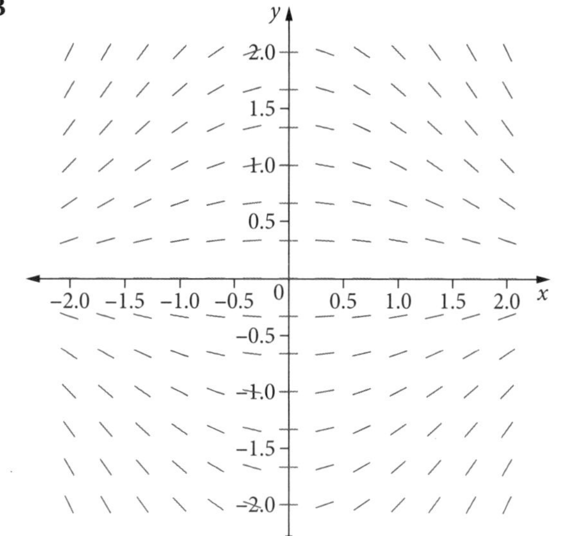

C

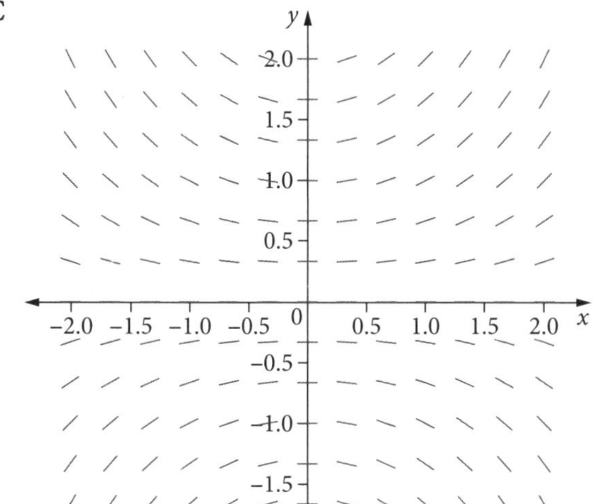

D

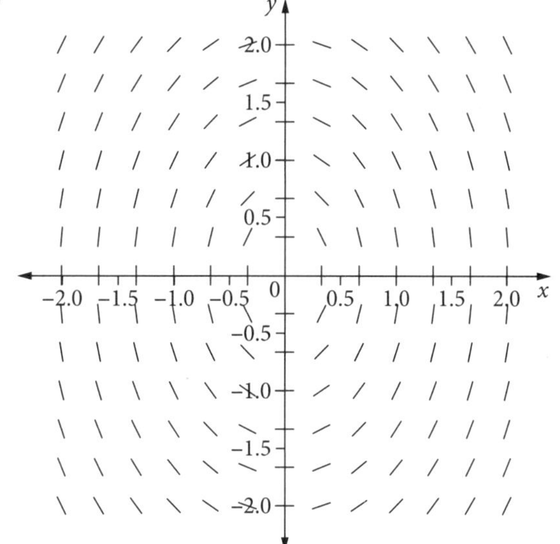

Question 12

Which equation satisfies the differential equation $y'' + 25y = 0$?

A $y = e^{-5x}$

B $y = e^{5x}$

C $y = 5\sin x$

D $y = \cos 5x$

Question 13

Which equation is a solution to the differential equation $\frac{dy}{dx} = -\frac{x}{y}$?

A $x^2 + y^2 = 9$

B $x^2 - y^2 = 25$

C $y = \sqrt{4 - (x-1)^2}$

D $y = \frac{1}{x}$

Question 14

For the differential equation $\frac{dy}{dx} = 0.1y(20 - y)$, for what value of y is the rate of change a maximum?

A $y = 0.1$

B $y = 5$

C $y = 10$

D $y = 20$

Question 15

The equation $\frac{dP}{dt} = 0.4P(5000 - P)$ is used to model the population of kangaroos in a national park.

What is the maximum population of kangaroos?

A 50

B 200

C 2500

D 5000

Question 16

Which of these differential equations has $y = 2e^{x^2}$ as a solution?

A $y'' = 4x^2y + 2y$

B $y' + 4xy = 0$

C $y'' = x^2y$

D $\frac{y''}{x} - 2y' = 0$

Question 17

Lucy performed the following steps to solve the differential equation $\frac{dy}{dx} = xy - 3x + 4y - 12$ and made mistakes.

In which line(s) did she make errors?

$$\frac{dy}{dx} = xy - 3x + 4y - 12$$

$$= (x-4)(y-3) \quad \text{Line 1}$$

$$\int \frac{1}{y-3}\,dy = \int \frac{1}{x-4}\,dx \quad \text{Line 2}$$

$$\ln|y-3| = \ln|x-4| + c \quad \text{Line 3}$$

A Lines 1 and 2

B Lines 1 and 3

C Line 2 only

D Lines 2 and 3

Question 18

Which function is a possible solution to $\frac{dy}{dx} = (2x - 2)(y + 1)$, given $y > 1$?

A $y = e^{-(x-1)^2}$

B $y = e^{x^2+2x} - 1$

C $y = e^{(x+1)^2} + 1$

D $y = e^{(x-1)^2}$

Question 19

Which direction field best matches the differential equation $\frac{dy}{dx} = \frac{x+y}{x-y}$?

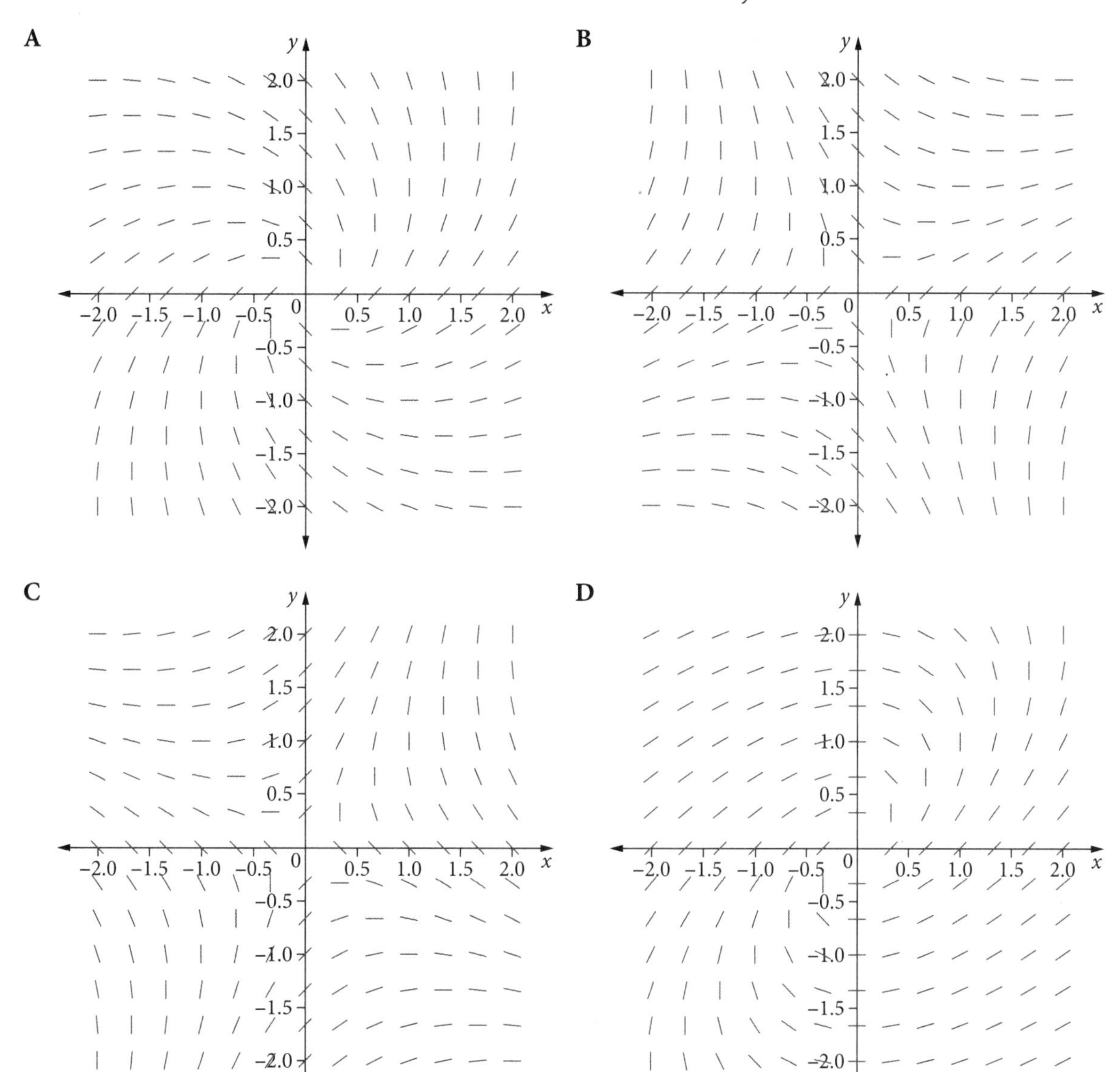

Question 20

An ice cube is removed from the freezer and placed on a plate. Initially, the temperature of the ice cube is 0°C.

Assuming the temperature of the ice cube can be modelled using $\frac{dT}{dt} = k(T - 25)$, which of the following equations correctly models the temperature of the ice cube?

A $T = 25 + 25e^{kt}$

B $T = 25 - 25e^{kt}$

C $T = 25 + 25e^{-kt}$

D $T = 25 - 25e^{-kt}$

Practice set 2

Short-answer questions

Solutions start on page 128.

Question 1 (2 marks)

Show that $y = Ae^{2x} + Be^{x}$ is a solution to the differential equation $y'' - 3y' + 2y = 0$. 2 marks

Question 2 (4 marks) ©NESA 2005 HSC EXAM, QUESTION 2(d)

A salad, which is initially at a temperature of 25°C, is placed in a refrigerator that has a constant temperature of 3°C. The cooling rate of the salad is proportional to the difference between the temperature of the refrigerator and the temperature, T, of the salad. That is, T satisfies the equation:

$$\frac{dT}{dt} = -k(T - 3),$$

where t is the number of minutes after the salad is placed in the refrigerator.

a Show that $T = 3 + Ae^{-kt}$ satisfies the equation. 1 mark

b The temperature of the salad is 11°C after 10 minutes. Find the temperature of the salad after 15 minutes, correct to one decimal place. 3 marks

Question 3 (2 marks) ©NESA 2020 HSC EXAM, QUESTION 11(e)

Solve $\frac{dy}{dx} = e^{2y}$, finding x as a function of y. 2 marks

Question 4 (2 marks)

Solve the differential equation $\frac{dy}{dx} = y^2 + 25$, given that the solution passes through the origin. 2 marks

Question 5 (2 marks)

Sketch a possible solution to the following direction field passing through the value $(1, 3)$, showing all relevant features. 2 marks

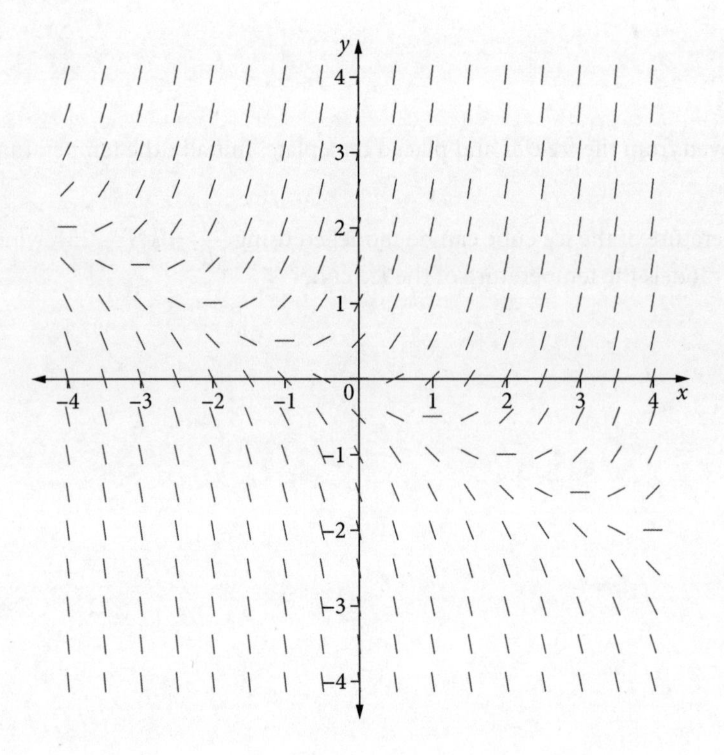

Question 6 (7 marks)

An object is put into a freezer to cool. After t minutes, its temperature is T°C. The freezer is at a constant temperature of -8°C.

The object's temperature decreases according to the differential equation $\frac{dT}{dt} = k(T+8)$.

a If the initial temperature is 40°C, solve the differential equation for T in terms of k and t. 3 marks

b If the object cools to 30°C in half an hour, show that $k = \frac{1}{30}\ln\left(\frac{19}{24}\right)$. 1 mark

c When will the temperature of the object be 0°C? Give your answer correct to the nearest minute. 2 marks

d Explain what will happen to T eventually. 1 mark

Question 7 (2 marks)

Solve the differential equation $\frac{dy}{dx} = x^2(1-y)$. 2 marks

Question 8 (2 marks)

Solve the differential equation $\frac{dy}{dx} = x - x\sin^2 y$, given that the graph of the solution passes through the point $(1, 0)$. 2 marks

Question 9 (2 marks)

Solve the differential equation $\frac{dy}{dx} = x\sqrt{9-y^2}$ when $y(1) = \frac{3\pi}{2}$. 2 marks

Question 10 (5 marks)

A town population of size P is known to have a rate of growth proportional to P.
In 1986, the town had a population of 1000 and in 1996, there were 1400 people.

a Determine the expected population in 2006. 3 marks

b In what year would the town expect to reach a population of 3000? 2 marks

Question 11 (5 marks)

Consider the differential equation $\frac{dy}{dx} = (y-5)(x+2)$.

a Determine the constant solution to the differential equation. 1 mark

b Use separation of variables to solve the differential equation. 2 marks

c Determine the unique solution if the graph of its solution passes through $(-1, 6)$. 2 marks

Question 12 (3 marks)

Consider the differential equation $\frac{dy}{dx} = e^x y$, where $y > 0$.

Using separation of variables, find the solution if its curve passes through $(0, e^2)$. 3 marks

Question 13 (3 marks)

a Show that $\frac{1}{y(1-y)} = \frac{1}{y} + \frac{1}{1-y}$. 1 mark

b Hence, solve the differential equation $\frac{dy}{dx} = y(1-y)$ if $0 < y < 1$ and the solution passes through $(0, 0.2)$. 2 marks

Question 14 (4 marks)

A pot of water was boiled to 100°C and left in a room with a constant temperature of 25°C. The rate of decrease of the temperature of the water T (°C), is proportional to the difference between the current temperature of water and room temperature; that is, $\frac{dT}{dt} = -k(T - 25)$. After 5 minutes, the water is 85°C.

How long will it take for the temperature of the water to reach 40°C, correct to the nearest minute? 4 marks

Question 15 (6 marks)

$y = f(x)$ has the derivative $\frac{dy}{dx} = \frac{4x + 3}{y}$.

a Solve the differential equation if $y = f(x)$ passes through the point $(1, 1)$. 3 marks

b Hence, sketch the graph of $y = f(x)$. 3 marks

Question 16 (7 marks)

Consider the differential equation $\frac{dy}{dx} = (3 - y)\cos x$. A solution to this differential equation passes through $(0, 1)$ and is shown on the direction field below.

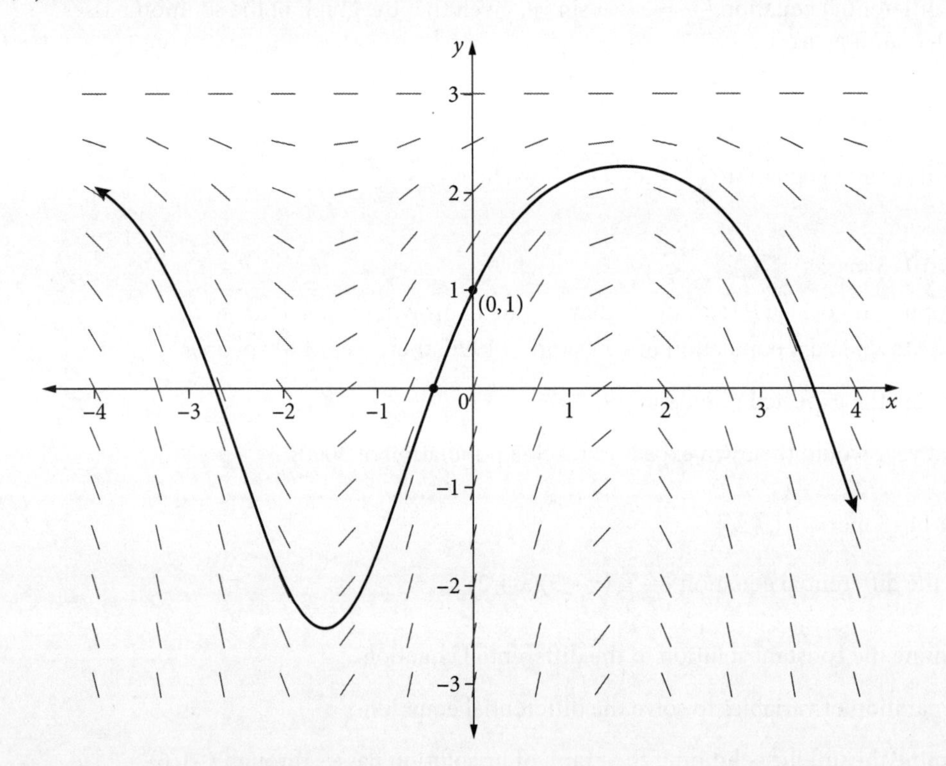

a Find the equation of this solution. 3 marks

b Find the equation of the tangent to the curve at $(0, 1)$, and find the x-intercept of this tangent. 2 marks

c Using parts **a** and **b**, find the distance between the tangent and the curve on the x-axis, correct to three decimal places. 2 marks

Question 17 (5 marks) ©NESA 2010 MATHEMATICS EXTENSION 2 HSC EXAM, QUESTION 5(c) MODIFIED

A TV channel has estimated that if it spends $\$x$ on advertising a particular program it will attract a proportion $y(x)$ of the potential audience for the program, where:

$$\frac{dy}{dx} = ay(1 - y)$$

and $a > 0$ is a given constant.

a Explain why $\frac{dy}{dx}$ has its maximum value when $y = \frac{1}{2}$. 1 mark

b Using the fact that $\int \frac{dy}{y(1 - y)} = \ln\left(\frac{y}{1 - y}\right) + c$, where $0 < y < 1$, deduce that: 3 marks

$$y(x) = \frac{1}{ke^{-ax} + 1}$$

for some constant $k > 0$.

c The TV channel knows that if it spends no money on advertising the program then the audience will be one-tenth of the potential audience. 1 mark

Find the value of the constant k referred to in part **b**.

Question 18 (6 marks)

The rate of decomposition of radium, a radioactive substance, is proportional to the amount present at any time t, $\frac{dQ}{dt} = -kQ$. The half-life of radium is 1600 years, that is, the time it takes for the original amount of radium to reduce to half.

a Show that $k = \frac{\ln 2}{1600}$. 2 marks

b What percentage of radium is remaining after 100 years? 1 mark

c After how many years will 80% of the mass be lost? Round up to the nearest year. 3 marks

Question 19 (8 marks)

A certain biological culture is modelled by the differential equation $\frac{dP}{dt} = 0.002P(100 - P)$, where $P(t)$ is the percentage of the petri dish covered by the culture at time t minutes. When first observed, the culture occupied 5% of the petri dish.

a Show that 1 mark

$$\frac{1}{0.002P(100 - P)} = 5\left(\frac{1}{P} + \frac{1}{100 - P}\right).$$

b Find the equation of the size, P, of the culture in terms of time t. 4 marks

c Find the percentage of the petri dish covered after 26 minutes. 1 mark

d Determine the time when the culture was growing at its maximum rate. 2 marks

PRACTICE SET 2

9780170459242

Question 20 (6 marks)

Police detectives determine an estimated time of death of a person by measuring the change in temperature of the body and applying Newton's law of cooling, which has the differential equation:

$$\frac{dT}{dt} = k(T - T_0),$$

where T is the temperature in °C, t is the time in hours after midnight and T_0 is the room temperature in °C.

Paul, a detective, arrived at the scene of the crime at midnight and measured the temperature of the body to be 26°C, where the room was at a constant temperature of 15°C.
The temperature of the body 3 hours later was 20°C.

a Show that $k \approx -0.2628$. 3 marks

b Hence, estimate the time of death, correct to the nearest minute, given that normal body temperature is 37°C. 3 marks

Practice set 1

Worked solutions

1 A

A: cannot be written in the form $\frac{dy}{dx} = f(x)g(y)$

B: $\frac{dy}{dx} = 4x - xy$
$= x(4 - y)$

C: $\frac{dy}{dx} = \sqrt{1 + x^2}$
$= 1\sqrt{1 + x^2}$

D: $\frac{dy}{dx} = 2^y \times 2^{-x}$

2 C

A: $\frac{dy}{dx} = \frac{1-2}{2+7}$
$= -\frac{1}{9}$

C: $\frac{dy}{dx} = \frac{1-(-1)}{2+7}$
$= \frac{2}{9}$

B: $\frac{dy}{dx} = \frac{1-(-3)}{2+(-11)}$
$= -\frac{4}{9}$

D: $\frac{dy}{dx} = \frac{1-2}{2+15}$
$= -\frac{1}{17}$

3 B

A: $\frac{dy}{dx} = 2x - 2$

$\frac{dy}{dx} + xy - x^3$
$= 2x - 2 + x(x^2 - 2x) - x^3$
$= 2x - 2 + x^3 - 2x^2 - x^3$
$= -2x^2 - 2x - 2$

B: $\frac{dy}{dx} = 2x$

$\frac{dy}{dx} + xy - x^3$
$= 2x + x(x^2 - 2) - x^3$
$= 2x + x^3 - 2x - x^3$
$= 0$

C: $\frac{dy}{dx} = 2x + 2$

$\frac{dy}{dx} + xy - x^3$
$= 2x + 2 + x(x^2 + 2x) - x^3$
$= 2x + 2 + x^3 + 2x^2 - x^3$
$= 2x^2 + 2x + 2$

D: $\frac{dy}{dx} = 2x$

$\frac{dy}{dx} + xy - x^3$
$= 2x + x(x^2 + 2) - x^3$
$= 2x + x^3 + 2x - x^3$
$= 4x$

4 A

$y = e^{-2x}$
$y' = -2e^{-2x}$
$y'' = 4e^{-2x}$

A: $y'' + 2y' = 4e^{-2x} - 4e^{-2x}$
$= 0$

B: $y'' + 2y' + 2y = 4e^{-2x} - 4e^{-2x} + 2e^{-2x}$
$= 2e^{-2x}$

C: $y' - 2y = -2e^{-2x} - 2e^{-2x}$
$= -4e^{-2x}$

D: $2y'' + 3y' + 2y = 8e^{-2x} - 6e^{-2x} + 2e^{-2x}$
$= 4e^{-2x}$

5 B

$\frac{dy}{dx} = 3x^2 - 7xy + 2y^2$
$= 3(2)^2 - 7(2)(5) + 2(5)^2$
$= -8$

6 B

Gradients are positive in the 1st quadrant (not C or D).

Gradients are negative in the 4th quadrant (not A).

7 B

The closest possible value the curve will pass through is $(-1.5, 0.5)$.

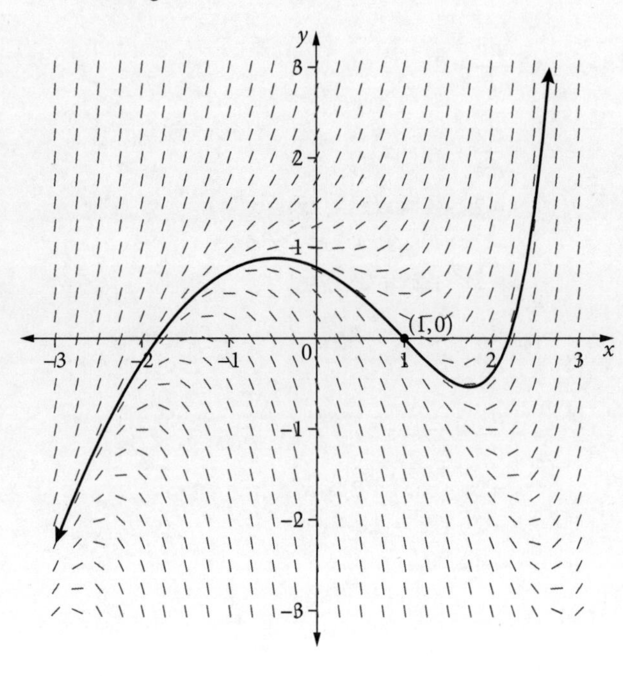

8 A

The gradients are all the same positive value. The general solution is linear with a positive gradient.

9 C

The gradients are steeper to the left and flatter to the right. The graph is exponential but $y \to 0$ as $x \to \infty$. The curve must have the term e^{-x}.

10 A

Gradients are positive in the first quadrant (not C or D).

Test gradient at $(-1, 1)$, which should be positive.

At A, $\frac{dy}{dx} = 3$. At B, $\frac{dy}{dx} = -1$.

11 C

When $y = 0$, $\frac{dy}{dx} = 0$. (not A or D)

When $x > 0$ and $y > 0$, $\frac{dy}{dx} > 0$. (not B)

12 D

A: $y = e^{-5x}$
$y' = -5e^{-5x}$
$y'' = 25e^{-5x}$

$y'' + 25y$
$= 25e^{-5x} + 25e^{-5x}$
$= 50e^{-5x}$

B: $y = e^{5x}$
$y' = 5e^{5x}$
$y'' = 25e^{5x}$

$y'' + 25y$
$= 25e^{5x} + 25e^{5x}$
$= 50e^{5x}$

C: $y = 5\sin x$
$y' = 5\cos x$
$y'' = -5\sin x$

$y'' + 25y$
$= -5\sin x + 125\sin x$
$= 120\sin x$

D: $y = \cos 5x$
$y' = -5\sin 5x$
$y'' = -25\cos 5x$

$y'' + 25y$
$= -25\cos 5x + 25\cos 5x$
$= 0$

13 A

$$\frac{dy}{dx} = -\frac{x}{y}$$

Using separation of variables:

$$\int y\,dy = \int -x\,dx$$
$$\frac{1}{2}y^2 = -\frac{1}{2}x^2 + c$$
$$y^2 = -x^2 + 2c$$
$$\therefore x^2 + y^2 = 2c$$

Then $x^2 + y^2 = 9$ is one solution to this differential equation.

14 C

The rate of change is the quadratic function $R = 0.1y(20 - y)$, whose graph is a concave-down parabola.

The zeroes of this parabola are $y = 0$ and $y = 20$.

The maximum value occurs at the vertex of the parabola, at the value of y halfway between 0 and 20, that is, the maximum rate of change is at $y = 10$.

15 D

Maximum and minimum population occurs when $\frac{dP}{dt} = 0$.

$$0.4P(5000 - P) = 0$$
$$P = 0 \text{ or } 5000$$

So maximum population is 5000.

16 A

$$y = 2e^{x^2}$$

$$y' = 4xe^{x^2}$$

$$y'' = 4x \times 2xe^{x^2} + 4e^{x^2}$$
$$= 8x^2e^{x^2} + 4e^{x^2}$$

A: $y'' = (4x^2 + 2)y$

$$(4x^2 + 2)y = (4x^2 + 2) \times 2e^{x^2}$$
$$= 8x^2e^{x^2} + 4e^{x^2}$$
$$= y''$$

B: $y' + 4xy = 4xe^{x^2} + 8xe^{x^2}$

$$= 12xe^{x^2} \neq 0$$

C: $x^2y = 2x^2e^{x^2} \neq y''$

D: $\frac{y''}{x} - 2y' = \frac{8x^2e^{x^2} + 4e^{x^2}}{x} - 8xe^{x^2}$

$$= 8xe^{x^2} + \frac{4}{x}e^{x^2} - 8xe^{x^2}$$
$$= \frac{4}{x}e^{x^2} \neq 0$$

17 A

Error in factorisation in Line 1.

Error in separating variables in Line 2.

Line 3 is correct.

Correct solution is:

$$\frac{dy}{dx} = xy - 3x + 4y - 12$$
$$= (x + 4)(y - 3)$$

$$\int \frac{1}{y-3} = \int x + 4\,dx$$

$$\ln|y - 3| = \frac{1}{2}x^2 + 4x + c$$

18 B

$$\frac{dy}{dx} = (2x - 2)(y + 1)$$

$$\int \frac{1}{y+1}\,dy = \int 2x - 2\,dx$$

$$\ln(y + 1) = x^2 + 2x + c \quad y > 1, \text{ so } y - 1 > 0$$

$$y + 1 = e^{x^2+2x+c}$$

$$y = e^{x^2+2x+c} - 1$$
$$= e^{x^2+2x} - 1 \text{ if } c = 0$$

An alternative solution is to test each of the answers, but that would take more time.

19 A

When $y = 0$, $\frac{dy}{dx} = 1$. (not C)

When $x = 0$, $\frac{dy}{dx} = -1$. (not D)

At $(1, 1)$, $\frac{dy}{dx} = \frac{1+1}{1-1}$, which is undefined (vertical gradient).

So answer is A.

20 B

$$\frac{dT}{dt} = k(T - 25)$$

$$\frac{dt}{dT} = \frac{1}{k(T-25)}$$

$$t = \frac{1}{k}\int \frac{1}{T-25}\,dT$$
$$= \frac{1}{k}\ln|T - 25| + c$$

When $t = 0$, $T = 0$:

$$0 = \frac{1}{k}\ln|0 - 25| + c$$

$$c = -\frac{1}{k}\ln 25$$

$$\therefore\ t = \frac{1}{k}\ln|T - 25| - \frac{1}{k}\ln 25$$

$$kt = \ln\left(\frac{25 - T}{25}\right) \quad (T < 25)$$

$$e^{kt} = \frac{25 - T}{25}$$

$$25e^{kt} = 25 - T$$

$$T = 25 - 25e^{kt}$$

WORKED SOLUTIONS

Practice set 2

Worked solutions

Question 1

$y = Ae^{2x} + Be^{x}$

$y' = 2Ae^{2x} + Be^{x}$

$y'' = 4Ae^{2x} + Be^{x}$

Then $y'' - 3y' + 2y$

$= 4Ae^{2x} + Be^{x} - 3(2Ae^{2x} + Be^{x}) + 2(Ae^{2x} + Be^{x})$

$= e^{2x}(4A - 6A + 2A) + e^{2x}(B - 3B + 2B)$

$= 0$

So $y'' - 3y' + 2y = 0$.

Question 2

a $T = 3 + Ae^{-kt}$

$$\begin{aligned}\frac{dT}{dt} &= -Ake^{-kt}\\ &= -k(Ae^{-kt})\\ &= -k(T - 3)\end{aligned}$$

b When $t = 0$, $T = 25$

$$\begin{aligned}25 &= 3 + Ae^{0}\\ A &= 25 - 3\\ &= 22\end{aligned}$$

So $T = 3 + 22e^{-kt}$.

When $t = 10$, $T = 11$:

$$\begin{aligned}11 &= 3 + 22e^{-10k}\\ 22e^{-10k} &= 8\\ e^{-10k} &= \frac{4}{11}\\ -10k &= \ln\left(\frac{4}{11}\right)\\ k &= -\frac{1}{10}\ln\left(\frac{4}{11}\right)\end{aligned}$$

So $T = 3 + 22e^{\frac{t}{10}\ln\left(\frac{4}{11}\right)}$.

At $t = 15$, $T = 3 + 22e^{\frac{15}{10}\ln\left(\frac{4}{11}\right)}$

$= 7.82$

$\approx 7.8°C$

Question 3

$$\frac{dy}{dx} = e^{2y}$$

$$\frac{dx}{dy} = e^{-2y}$$

$$\begin{aligned}x &= \int e^{-2y}\,dy\\ &= -\frac{1}{2}e^{-2y} + c\end{aligned}$$

Question 4

$$\frac{dy}{dx} = y^2 + 25$$

$$\frac{dx}{dy} = \frac{1}{y^2 + 25}$$

$$\begin{aligned}x &= \int \frac{1}{y^2 + 25}\,dy\\ &= \tan^{-1}\left(\frac{y}{5}\right) + c\end{aligned}$$

When $x = 0$, $y = 0$:

$0 = \tan^{-1}(0) + c$

$c = 0$

So $x = \tan^{-1}\left(\frac{y}{5}\right)$.

$$\frac{y}{5} = \tan x$$

$$y = 5\tan x$$

Question 5

The actual solution is as follows. With only a direction field, accuracy in the solution is difficult. However, as a guide, an ideal solution should include the following key points:

- The graph should pass through the point $(1, 3)$.
- The graph should turn near $x = 0$ but above (or just touching) the x-axis.
- As $x \to -\infty$, the graph should increase and approach a straight line.

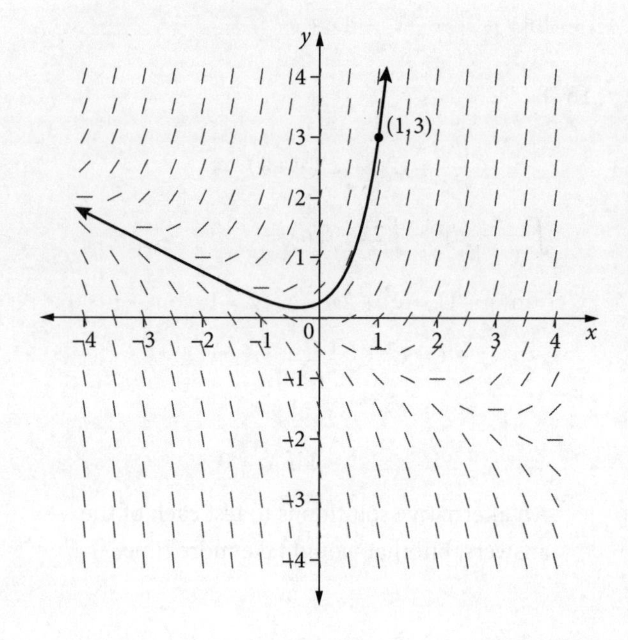

Question 6

a $\dfrac{dT}{dt} = k(T+8)$

$$\frac{dt}{dT} = \frac{1}{k(T+8)}$$

$$t = \frac{1}{k}\int \frac{1}{T+8}\,dT$$

$$= \frac{1}{k}\ln(T+8) + c \quad (T > -8)$$

When $t = 0$, $T = 40$:

$$0 = \frac{1}{k}\ln(40+8) + c$$

$$c = -\frac{1}{k}\ln 48$$

$$\therefore t = \frac{1}{k}(\ln(T+8) - \ln 48)$$

$$kt = \ln\left(\frac{T+8}{48}\right)$$

$$\frac{T+8}{48} = e^{kt}$$

$$T + 8 = 48e^{kt}$$

$$T = 48e^{kt} - 8$$

b When $t = 30$, $T = 30$:

$$30 = 48e^{30k} - 8$$

$$48e^{30k} = 38$$

$$e^{-30k} = \frac{38}{48} = \frac{19}{24}$$

$$30k = \ln\left(\frac{19}{24}\right)$$

$$k = \frac{1}{30}\ln\left(\frac{19}{24}\right)$$

c So $T = 48e^{\frac{t}{30}\ln\left(\frac{19}{24}\right)} - 8$

or $T = 48\left(\frac{19}{24}\right)^{\frac{t}{30}} - 8$.

When $T = 0$:

$$0 = 48e^{\frac{t}{30}\ln\frac{19}{24}} - 8$$

$$48e^{\frac{t}{30}\ln\frac{19}{24}} = 8$$

$$e^{\frac{t}{30}\ln\frac{19}{24}} = \frac{8}{48} = \frac{1}{6}$$

$$\frac{t}{30}\ln\left(\frac{19}{24}\right) = \ln\left(\frac{1}{6}\right)$$

$$t = \frac{30\ln\left(\frac{1}{6}\right)}{\ln\left(\frac{19}{24}\right)}$$

$$= 230.1$$

$$\approx 230 \text{ minutes}$$

d As $t \to \infty$, e^{kt} or $\left(\frac{19}{24}\right)^{\frac{t}{30}} \to 0$ (since $k < 0$),

$T \to 48 \times 0 - 8 = -8$

So $T \to -8$.

Question 7

$$\frac{dy}{dx} = x^2(1-y)$$

$$\int \frac{1}{1-y}\,dy = \int x^2\,dx$$

$$-\ln|1-y| = \frac{1}{3}x^3 + c$$

$$\ln|1-y| = -\frac{1}{3}x^3 - c$$

$$|1-y| = e^{-\frac{1}{3}x^3 - c}$$

$$= Ae^{-\frac{1}{3}x^3} \quad \text{where } A = e^{-c} \text{ is a constant}$$

$$1 - y = Be^{-\frac{1}{3}x^3} \quad \text{where } B \text{ can be a positive or negative constant } (A \text{ or } -A)$$

So $y = 1 - Be^{-\frac{1}{3}x^3}$.

Question 8

$$\frac{dy}{dx} = x(1 - \sin^2 y)$$

$$= x\cos^2 y$$

$$\int \sec^2 y\,dy = \int x\,dx$$

$$\tan y = \frac{1}{2}x^2 + c$$

When $x = 1$, $y = 0$:

$$\tan 0 = \frac{1}{2} + c$$

$$c = -\frac{1}{2}$$

$$\therefore \tan y = \frac{1}{2}(x^2 - 1)$$

$$y = \tan^{-1}\left(\frac{1}{2}(x^2 - 1)\right)$$

Question 9

$$\frac{dy}{dx} = 2x\sqrt{9 - y^2}$$

$$\int \frac{1}{\sqrt{9-y^2}}\,dy = \int 2x\,dx$$

$$\sin^{-1}\left(\frac{y}{3}\right) = x^2 + c$$

When $x = 1$, $y = \frac{3\pi}{2}$:

$$\sin^{-1}\left(\frac{\pi}{2}\right) = 1^2 + c$$

$$c = 1 - 1$$

$$= 0$$

$$\therefore \sin^{-1}\left(\frac{y}{3}\right) = x^2$$

$$\frac{y}{3} = \sin x^2$$

$$y = 3\sin x^2$$

Question 10

a $\frac{dP}{dt} = kP$

$\frac{dt}{dP} = \frac{1}{kP}$

$t = \frac{1}{k}\int \frac{1}{P}\,dP$

$= \frac{1}{k}\ln P + c \qquad (P > 0)$

Let t = number of years after 1986.

When $t = 0$, $P = 1000$:

$0 = \frac{1}{k}\ln 1000 + c$

$c = -\frac{1}{k}\ln 1000$

$t = \frac{1}{k}(\ln P - \ln 1000)$

$= \frac{1}{k}\ln\frac{P}{1000}$

$kt = \ln\frac{P}{1000}$

$\frac{P}{1000} = e^{kt}$

So $P = 1000e^{kt}$.

When $t = 10$, $P = 1400$:

$1400 = 1000e^{10k}$

$e^{10k} = 1.4$

$10k = \ln 1.4$

$k = \frac{1}{10}\ln 1.4$

So $P = 1000e^{0.1t\ln 1.4}$

or $P = 1000(1.4)^{0.1t}$

At $t = 20$:

So $P = 1000e^{\frac{1}{10}\times 20\ln 1.4}$

≈ 1960 people.

b When $P = 3000$:

$3000 = 1000e^{\frac{1}{10}t\ln 1.4}$

$3 = e^{\frac{1}{10}t\ln 1.4}$

$\frac{1}{10}t\ln 1.4 = \ln 3$

$t = \frac{\ln 3}{\frac{1}{10}\ln 1.4}$

$= 32.65$

≈ 33 years

The town will reach a population of 3000 in 2019.

Question 11

a The constant solution is when $g(y) = 0$.

$y - 5 = 0$

So $y = 5$.

b $\frac{dy}{dx} = (y-5)(x+2)$

$\int \frac{1}{y-5}\,dy = \int x + 2\,dx$

$\ln|y-5| = \frac{1}{2}x^2 + 2x + c$

$|y-5| = e^{\frac{1}{2}x^2+2x+c}$

$|y-5| = Ae^{\frac{1}{2}x^2+2x}$ where A is a constant

$y - 5 = Ae^{\frac{1}{2}x^2+2x}$ (A could be positive or negativ

$y = Ae^{\frac{1}{2}x^2+2x} + 5$

c When $x = -1$, $y = 6$:

$6 = Ae^{\frac{1}{2}(-1)^2+2(-1)} + 5$

$1 = Ae^{-\frac{3}{2}}$

$A = e^{\frac{3}{2}}$

So $y = e^{\frac{3}{2}} \times e^{\frac{1}{2}x^2+2x} + 5$

$= e^{\frac{1}{2}x^2+2x+\frac{3}{2}} + 5.$

Question 12

$\frac{dy}{dx} = e^x y$

$\int \frac{1}{y}\,dy = \int e^x\,dx$

$\ln y = e^x + c \quad (y > 0)$

When $x = 0$, $y = e^2$:

$\ln e^2 = e^0 + c$

$2 = 1 + c$

$c = 1$

So $\ln y = e^x + 1$

$y = e^{e^x+1}.$

Question 13

a $\text{RHS} = \frac{1}{y} + \frac{1}{1-y}$

$= \frac{1-y+y}{y(1-y)}$

$= \frac{1}{y(1-y)}$

$= \text{LHS}$

b $\frac{dy}{dx} = y(1-y)$

$\frac{dx}{dy} = \frac{1}{y(1-y)}$

$x = \int \frac{1}{y(1-y)}\,dy$

$= \int \frac{1}{y} + \frac{1}{1-y}\,dy$

$= \ln(y) - \ln(1-y) + c \quad (0 < y < 1)$

$= \ln\left(\frac{y}{1-y}\right) + c$

When $x = 0$, $y = 0.2$:

$0 = \ln\left(\frac{0.2}{0.8}\right) + c$

$c = -\ln\frac{1}{4}$

$x = \ln\left(\frac{y}{1-y}\right) - \ln\frac{1}{4}$

$= \ln\left(\frac{y}{\frac{1}{4}(1-y)}\right)$

$= \ln\left(\frac{4y}{1-y}\right)$

$e^x = \frac{4y}{1-y}$

$4y = e^x(1-y)$

$= e^x - ye^x$

$y(4+e^x) = e^x$

So $y = \frac{e^x}{4+e^x}$.

Question 14

$\frac{dT}{dt} = -k(T-25)$

$\frac{dt}{dT} = -\frac{1}{k(T-25)}$

$t = -\frac{1}{k}\int \frac{1}{T-25}\,dT$

$= -\frac{1}{k}\ln(T-25) + c \quad (T > 25)$

When $t = 0$, $T = 100$:

$0 = -\frac{1}{k}\ln(100-25) + c$

$c = \frac{1}{k}\ln 75$

$\therefore t = -\frac{1}{k}(\ln(T-25) - \ln 75)$

$= -\frac{1}{k}\ln\left(\frac{T-25}{75}\right)$

$-kt = \ln\left(\frac{T-25}{75}\right)$

$\frac{T-25}{75} = e^{-kt}$

$T - 25 = 75e^{-kt}$

$T = 75e^{-kt} + 25$

When $t = 5$, $T = 85$:

$85 = 75e^{-5k} + 25$

$60 = 75e^{-5k}$

$-5k = \ln\frac{60}{75}$

$k = -\frac{1}{5}\ln\frac{4}{5}$

So $T = 75e^{\frac{1}{5}t\ln\frac{4}{5}} + 25$.

If $T = 40$:

$40 = 75e^{\frac{1}{5}t\ln\frac{4}{5}} + 25$

$75e^{\frac{1}{5}t\ln\frac{4}{5}} = 15$

$e^{\frac{1}{5}t\ln\frac{4}{5}} = \frac{1}{5}$

$\frac{1}{5}t\ln\frac{4}{5} = \ln\frac{1}{5}$

$t = \frac{5\ln\frac{1}{5}}{\ln\frac{4}{5}}$

$= 36.1$

≈ 36 minutes

Question 15

a $\dfrac{dy}{dx} = \dfrac{4x+3}{y}$

$\int y\,dy = \int 4x + 3\,dx$

$\frac{1}{2}y^2 = 2x^2 + 3x + c$

When $x = 1, y = 1$:

$$\frac{1}{2} = 2 + 3 + c$$

$$c = -\frac{9}{2}$$

$$\therefore \frac{1}{2}y^2 = 2x^2 + 3x - \frac{9}{2}$$

$$y^2 = 4x^2 + 6x - 9$$

$$y = \pm\sqrt{4x^2 + 6x - 9}$$

$$= \sqrt{4x^2 + 6x - 9} \qquad (y(1) = 1)$$

b Note: This part of the question is a crossover with the Year 11 Further graphing topic.

Find the roots of the parabola.

$4x^2 + 6x - 9 = 0$

$$x = \frac{-6 \pm \sqrt{6^2 - 4(4)(-9)}}{2(4)}$$

$$= \frac{-6 \pm \sqrt{180}}{8}$$

$$= \frac{-6 \pm 6\sqrt{5}}{8}$$

$$= \frac{-3 \pm 3\sqrt{5}}{4}$$

$$\approx -2.427, 0.927$$

Hence, sketch the graph of the square root from the graph of the parabola.

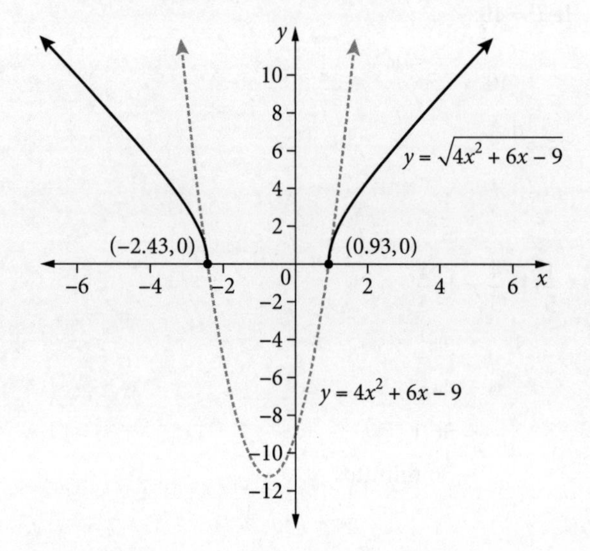

Question 16

a $\dfrac{dy}{dx} = (3-y)\cos x$

$\int \frac{1}{3-y}\,dy = \int \cos x\,dx$

$-\ln|3-y| = \sin x + c$

As $y < 3$ from the graph, the absolute value can be removed.

$-\ln(3-y) = \sin x + c$

When $x = 0, y = 1$:

$$-\ln(3-1) = \sin 0 + c$$

$$c = -\ln 2$$

$$\therefore -\ln(3-y) = \sin x - \ln 2$$

$$\ln\frac{2}{3-y} = \sin x$$

$$\frac{2}{3-y} = e^{\sin x}$$

$$3 - y = 2e^{-\sin x}$$

So $y = 3 - 2e^{-\sin x}$.

b At $(0, 1)$:

$$\frac{dy}{dx} = (3-y)\cos x$$

$$= (3-1)\cos 0$$

$$= 2$$

Then equation of tangent is:

$$y - 1 = 2(x - 0)$$

$$y = 2x + 1$$

x-intercept of tangent when $y = 0$:

$$2x + 1 = 0$$

$$2x = -1$$

$$x = -\frac{1}{2}$$

c x-intercept of curve when $y = 0$:

$$0 = 3 - 2e^{-\sin x}$$

$$2e^{-\sin x} = 3$$

$$e^{-\sin x} = \frac{3}{2}$$

$$-\sin x = \ln\left(\frac{3}{2}\right)$$

$$\sin x = -\ln\left(\frac{3}{2}\right)$$

$$x = -0.417$$ since we want the negative solution closest to 0

Distance $= -0.417 - (-0.5)$

$= 0.083$ units

Question 17

a $R = ay(1 - y)$

The rate R can be represented by a parabola with roots $y = 0$, $y = 1$.

The maximum value will occur halfway between 0 and 1.

So $\frac{dy}{dx}$ is maximised at $y = \frac{1}{2}$.

b $\frac{dy}{dx} = ay(1 - y)$

$$\frac{dx}{dy} = \frac{1}{ay(1-y)}$$

$$x = \frac{1}{a}\int \frac{1}{y(1-y)}\,dy$$

$$= \frac{1}{a}\ln\left(\frac{y}{1-y}\right) + c$$

$$x - c = \frac{1}{a}\ln\left(\frac{y}{1-y}\right)$$

$$a(x - c) = \ln\left(\frac{y}{1-y}\right)$$

$$\frac{y}{1-y} = e^{a(x-c)}$$

$$\frac{1-y}{y} = e^{-ax+ac}$$

$$\frac{1}{y} - 1 = ke^{-ax} \text{ for a constant } k = e^{ac}$$

$$\frac{1}{y} = ke^{-ax} + 1$$

$$\text{So } y = \frac{1}{ke^{-ax} + 1}.$$

c When $x = 0$, $y = 0.1$:

$$0.1 = \frac{1}{ke^{0} + 1}$$

$$0.1 = \frac{1}{k + 1}$$

$$k + 1 = 10$$

$$k = 9$$

Question 18

a $\frac{dQ}{dt} = -kQ$

$$\frac{dt}{dQ} = -\frac{1}{kQ}$$

$$t = -\frac{1}{k}\int \frac{1}{Q}\,dQ$$

$$= -\frac{1}{k}\ln Q + c \quad (Q > 0)$$

Let Q_0 be the original quantity.

When $t = 0$, $Q = Q_0$:

$$0 = -\frac{1}{k}\ln Q_0 + c$$

$$c = \frac{1}{k}\ln Q_0$$

$$t = -\frac{1}{k}(\ln Q - \ln Q_0)$$

$$= -\frac{1}{k}\ln\left(\frac{Q}{Q_0}\right)$$

$$-kt = \ln\left(\frac{Q}{Q_0}\right)$$

$$\frac{Q}{Q_0} = e^{-kt}$$

$$Q = Q_0e^{-kt}$$

When $t = 1600$, $Q = \frac{1}{2}Q_0$:

$$\frac{1}{2}Q_0 = Q_0e^{-1600k}$$

$$e^{-1600k} = \frac{1}{2}$$

$$-1600k = \ln\left(\frac{1}{2}\right)$$

$$= \ln 2^{-1}$$

$$= -\ln 2$$

$$\text{So } k = \frac{\ln 2}{1600}.$$

b When $t = 100$:

$$Q = Q_0e^{-kt}$$

$$= Q_0e^{-100\times\frac{\ln 2}{1600}}$$

$$= Q_0 \times 0.957\ldots$$

96% of the original radium remains.

c If 80% of the mass has been lost, 20% remains.

When $Q = 0.2Q_0$:

$$0.2Q_0 = Q_0e^{-\frac{\ln 2}{1600}t}$$

$$e^{-\frac{\ln 2}{1600}t} = 0.2$$

$$-\frac{\ln 2}{1600}t = \ln 0.2$$

$$t = -\frac{1600\ln 0.2}{\ln 2}$$

$$= 3715.1$$

$$\approx 3716 \text{ years (rounding up)}$$

Question 19

a $\text{RHS} = 5\left(\frac{1}{P} + \frac{1}{100 - P}\right)$

$= 5\left(\frac{100 - P + P}{P(100 - P)}\right)$

$= \frac{500}{P(100 - P)}$

$= \frac{1}{0.002P(100 - P)}$

$= \text{LHS}$

b $\frac{dP}{dt} = 0.002P(100 - P)$

$\frac{dt}{dP} = \frac{1}{0.002P(100 - P)}$

$t = \int \frac{1}{0.002P(100 - P)}\, dP$

$= 5\int \frac{1}{P} + \frac{1}{100 - P}\, dP$ (from part **a**)

$= 5\left(\ln|P| - \ln|100 - P|\right) + c$

$= 5\ln\left(\frac{P}{100 - P}\right) + c \quad (0 < P < 100)$

When $t = 0$, $P = 5$:

$0 = 5\ln\left(\frac{5}{100 - 5}\right) + c$

$c = -5\ln\left(\frac{1}{19}\right)$

$= 5\ln 19$

$\therefore t = 5\ln\left(\frac{P}{100 - P}\right) + 5\ln 19$

$= 5\ln\left(\frac{19P}{100 - P}\right)$

$\frac{2}{10}t = \ln\left(\frac{19P}{100 - P}\right)$

$\frac{19P}{100 - P} = e^{\frac{2}{10}t}$

$\frac{100 - P}{19P} = e^{-\frac{2}{10}t}$

$100 - P = 19Pe^{-\frac{2}{10}t}$

$100 = P(19e^{-\frac{2}{10}t} + 1)$

So $P = \frac{100}{19e^{-\frac{2}{10}t} + 1}$.

c When $t = 26$:

$P = \frac{100}{19e^{-0.2\times 26} + 1}$

$= 90.51$

$\approx 91\%$

d The graph of $\frac{dP}{dt}$ is shown below, with P-intercepts at 0 and 100. So the maximum occurs halfway between 0 and 100.

Maximum rate occurs when $P = 50$:

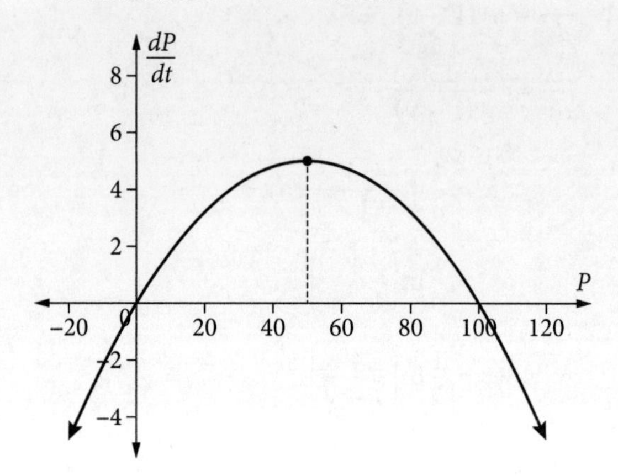

When $P = 50$:

$50 = \frac{100}{19e^{-\frac{2}{10}t} + 1}$

$19e^{-\frac{2}{10}t} + 1 = 2$

$19e^{-\frac{2}{10}t} = 1$

$e^{-\frac{2}{10}t} = \frac{1}{19}$

$-\frac{2}{10}t = \ln\left(\frac{1}{19}\right)$

$t = -5\ln\left(\frac{1}{19}\right)$

$= 5\ln 19$

$= 14.722\ldots$

≈ 15 minutes

Question 20

a $\dfrac{dT}{dt} = k(T-15)$

$$\frac{dt}{dT} = \frac{1}{k(T-15)}$$

$$t = \frac{1}{k}\int \frac{1}{T-15}\,dT$$

$$kt = \ln(T-15) + c \qquad (T > 15)$$

When $t = 0$, $T = 26$:

$$0 = \ln(26-15) + c$$
$$c = -\ln 11$$

$$\therefore kt = \ln(T-15) - \ln 11$$
$$= \ln\left(\frac{T-15}{11}\right)$$
$$\frac{T-15}{11} = e^{kt}$$
$$\therefore T = 11e^{kt} + 15$$

When $t = 3$, $T = 20$:

$$20 = 11e^{3k} + 15$$
$$11e^{3k} = 5$$
$$e^{3k} = \frac{5}{11}$$
$$3k = \ln\left(\frac{5}{11}\right)$$
$$k = \frac{1}{3}\ln\left(\frac{5}{11}\right)$$
$$\approx -0.2628$$

b When $T = 37$:

$$37 = 11e^{kt} + 15$$
$$11e^{kt} = 22$$
$$e^{kt} = 2$$
$$kt = \ln 2$$
$$-0.2628t = \ln 2$$
$$t = -\frac{\ln 2}{0.2628}$$
$$= -2.6375$$
$$= -2 \text{ hours and } 38 \text{ minutes}$$

The body was 37°C 2 hours and 38 minutes before midnight.

So the time of death was 9.22 pm.

WORKED SOLUTIONS

HSC exam topic grid (2011–2020)

This table shows the coverage of this topic in past HSC exams by question number. The past exams can be downloaded from the NESA website (www.educationstandards.nsw.edu.au) by selecting 'Year 11 – Year 12', 'HSC exam papers'. NESA marking feedback and guidelines can also be found there.

The new Mathematics Extension 1 course was first examined in 2020. For exams before 2020, select 'Year 11 – Year 12', 'Resources archive', 'HSC exam papers archive'.

Differential equations were introduced to the Mathematics Extension 1 course in 2020, although similar questions have been asked in previous years.

	Solving differential equations	Direction fields	Applications of differential equations	Exponential growth and decay
2011	4(c) Maths Extension 2 exam			5(b)
2012				
2013				12(c)
2014				12(f)
2015	2			
2016				12(b)
2017			14(c)	
2018				5
2019	12(d)			12(d)
2020 new course	**11(e)**, 12(e)	7		

The question in **bold** can be found on page 120 in this chapter.

CHAPTER 6
THE BINOMIAL DISTRIBUTION

ME-S1 The binomial distribution 140
S1.1 Bernoulli and binomial distributions 140
S1.2 Normal approximation for the sample proportion 142

THE BINOMIAL DISTRIBUTION

Binomial distribution

- Bernoulli distributions
- Binomial distributions
- Mean and variance

$$E(X) = np$$
$$\mathrm{Var}(X) = np(1 - p)$$

Binomial probability

$$P(X = r) = {}^nC_r p^r (1 - p)^{n - r}$$
$$X \sim \mathrm{Bin}(n, p)$$

Sample proportions

- Sample proportion distributions

$$E(\hat{p}) = p$$
$$\mathrm{Var}(\hat{p}) = \frac{pq}{n}$$

- Normal approximation

GLOSSARY

A+ DIGITAL FLASHCARDS
Revise this topic's key terms and concepts by scanning the QR code or typing the URL into your browser.

https://get.ga/a-hsc-maths-ext-1

Bernoulli distribution
A discrete probability distribution that has only 2 possible outcomes: 1 for success with probability p, or 0 for failure with probability $q = 1 - p$.

Bernoulli random variable
A discrete random variable that can take the value 0 or 1.

Bernoulli trial
An experiment with only 2 possible outcomes: success or failure.

binomial distribution
A discrete probability distribution of the number of successes achieved when n identical, independent Bernoulli trials are performed.

$$P(X = r) = {}^nC_r p^r (1 - p)^{n-r}$$
$$E(X) = np$$
$$\text{Var}(X) = np(1 - p)$$

binomial random variable
A discrete random variable that counts the number of successes in n Bernoulli trials.

$$X \sim \text{Bin}(n, p)$$

mean
The expected value of a random variable as an experiment is repeated.

normal approximation for the sample proportion
A binomial distribution can be approximated using a normal distribution if $np \geq 5$ and $nq \geq 5$ where $q = 1 - p$.

sample proportion
The proportion or fraction of successful Bernoulli trials out of the total number of trials taken.

$$\hat{p} = \frac{x}{n}$$
$$E(\hat{p}) = p$$
$$\text{Var}(\hat{p}) = \frac{pq}{n}$$

standard deviation
A measure of the spread of a probability distribution. The square root of the variance.

variance
A measure of the spread of a probability distribution. The square of the standard deviation.

GLOSSARY

Topic summary

The binomial distribution (ME-S1)

S1.1 Bernoulli and binomial distributions

Bernoulli distribution

A **Bernoulli random variable** is a discrete random variable with 2 possible outcomes: success and failure. If X is a Bernoulli random variable, the probability distribution can be represented by $P(X = x)$ for all x. X is either 0 (failure) or 1 (success). The probability of a success is p and the probability of a failure is $q = 1 - p$.

$$P(X = x) = \begin{cases} p & \text{for } x = 1 \\ 1 - p & \text{for } x = 0 \end{cases}$$

For example, rolling a 5 on a six-sided die is a **Bernoulli trial** where $p = \frac{1}{6}$.

The probability of **not** rolling a 5 (1, 2, 3, 4, 6) is a Bernoulli trial with a probability of $q = 1 - p = \frac{5}{6}$.

Binomial distribution

If a Bernoulli trial is repeated a number of times, then the **binomial random variable** is another discrete random variable that counts *the number of* successes over those trials. If there are n trials, then the number of successes, X, is a whole number from 0 to n.

For example, rolling a die 100 times and counting the frequency of 5 is a binomial trial.

Probability of success is p and probability of failure is q for a chance experiment with n independent, identical trials.

To find the probability of r successes:

$$P(X = r) = {}^nC_r p^r (1 - p)^{n-r}$$

$$\text{OR} \quad P(X = r) = {}^nC_r p^r q^{n-r}, \text{ where } q = 1 - p.$$

For example: Number of 5s in 100 rolls of a die.

$$\text{Probability of 13 5s: } P(X = 13) = {}^{100}C_{13}\left(\frac{1}{6}\right)^{13}\left(\frac{5}{6}\right)^{87}$$

$$\text{Probability of 20 5s: } P(X = 20) = {}^{100}C_{20}\left(\frac{1}{6}\right)^{20}\left(\frac{5}{6}\right)^{80}$$

$$\text{Probability of 56 5s: } P(X = 56) = {}^{100}C_{56}\left(\frac{1}{6}\right)^{56}\left(\frac{5}{6}\right)^{44}$$

$$\text{Probability of 75 5s: } P(X = 75) = {}^{100}C_{75}\left(\frac{1}{6}\right)^{75}\left(\frac{5}{6}\right)^{25}$$

A **binomial distribution** has n independent trials and a probability of success, p. A binomial random variable is written as $X \sim \text{Bin}(n, p)$ (can also be written as $\text{Bn}(n,p)$ or $\text{Bi}(n,p)$).

$$X \sim \text{Bin}(n,p) \Rightarrow P(X = x) = \binom{n}{x} p^x (1 - p)^{n-x}, x = 0, 1, \ldots, n$$

Hence, a Bernoulli random variable is $X \sim \text{Bin}(1, p)$. If $n = 1$, the trial in a binomial distribution is a Bernoulli random variable with only 2 outcomes: success (p) or failure ($1 - p$).

> **Hint**
> In binomial probability problems, first determine p, the probability of a success. It may be a compound event.

	Bernoulli distribution	**Binomial distribution**
Expected value (mean)	$E(X) = p$	$E(X) = np$
Variance	$\text{Var}(X) = pq$ $= p(1 - p)$	$\text{Var}(X) = npq$ $= np(1 - p)$
Standard deviation	$\sigma = \sqrt{pq}$	$\sigma = \sqrt{npq}$

> **Hint**
> The HSC exam reference sheet includes the binomial distribution formulas for mean and variance. Make sure you check it rather than relying on your memory in an exam.

- If $p = \frac{1}{2}$ (even chance), the graph of a binomial distribution is **symmetrical**.

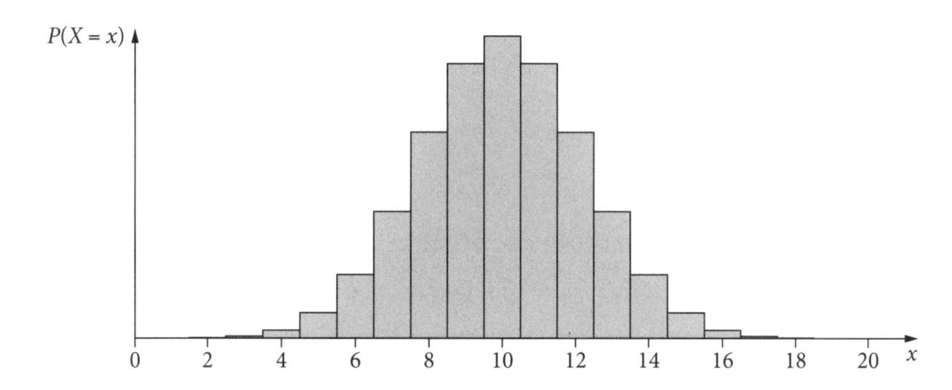

- If $p < \frac{1}{2}$ (unlikely), the **mean** is closer to the left and the graph is **positively skewed**.

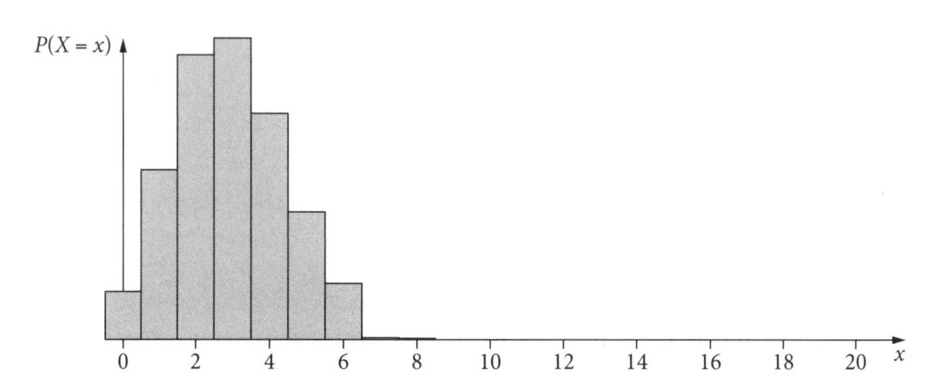

- If $p > \frac{1}{2}$ (likely), the mean is closer to the right and the graph is **negatively skewed**.

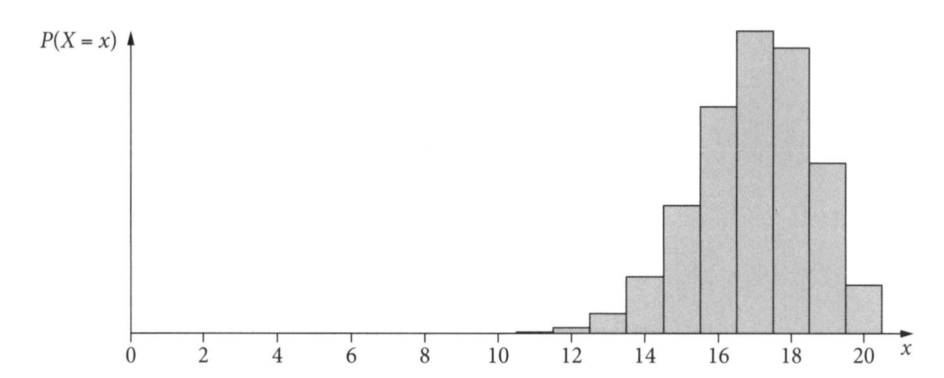

S1.2 Normal approximation for the sample proportion

Sample proportions

A **population proportion** p is the fraction of people or objects in the population that have a certain characteristic, such as left-handedness.

When the population is large, we usually take a large sample from the population and assume that the **sample proportion** $\hat{p}$ is a good **estimate** of the population proportion. The symbol ^ above p indicates that the sample proportion $\hat{p}$ is an estimate of the population proportion p. The sample proportion is written as

$$\hat{p} = \frac{x}{n},$$

where x is a value of a binomial random variable X with probability of success p and sample size n. Because X is a discrete random variable, $\hat{p} = \frac{x}{n}$ is also a discrete random variable. The value of $\hat{p} = \frac{x}{n}$ is different for each sample selected randomly from the population, so it has a discrete probability distribution called the **sampling distribution of proportions**, with mean and **variance**:

$$E(\hat{p}) = p \qquad \mathrm{Var}(\hat{p}) = \frac{pq}{n}$$

The mean of the sample proportion provides an estimate of the probability of a success in the Bernoulli trial.

Normal approximation for the sample proportion

A normal distribution will give an accurate approximation for the binomial distribution if the probability is close to $\frac{1}{2}$ or a large number of trials are conducted. As the sample size becomes larger, the sample distributions become closer to a normal distribution. This is the central limit theory. So a normal distribution can be used to approximate discrete data with large sample sizes as well as continuous data. This makes it easier to calculate the probabilities of a large range of possible outcomes, such as $P(X > 50)$ if $n = 100$. If treated as discrete data the time-consuming calculation would involve $P(X = 50, 51, 52, \ldots)$. Using the normal approximation, $P(X > 50)$ can be calculated simply using the z-score and a standard normal distribution table.

As a rule of thumb, a binomial distribution can be approximated using a normal distribution if $np \geq 5$ and $nq \geq 5$, where $q = 1 - p$.

The z-scores can be calculated for the section of the distribution required to find the probability once the mean and variance have been calculated.

Note: Remember to use the table of standard normal distribution values if it is given to you, or use the empirical rule on the HSC exam reference sheet if the z-scores are integers from -3 to 3.

Practice sets tracking grid

Maths is all about repetition, meaning do, do and do again! Each question in the following practice sets, especially the struggle questions (different for everybody!), should be completed at least 3 times correctly. Below is a tracking grid to record your question attempts: ✓ if you answered correctly, ✘ if you didn't.

PRACTICE SET 1: Multiple-choice questions

Question	1st attempt	2nd attempt	3rd attempt	4th attempt	5th attempt
1					
2					
3					
4					
5					
6					
7					
8					
9					
10					
11					
12					
13					
14					
15					
16					
17					
18					
19					
20					

PRACTICE SET 2: Short-answer questions

Question	1st attempt	2nd attempt	3rd attempt	4th attempt	5th attempt
1					
2					
3					
4					
5					
6					
7					
8					
9					
10					
11					
12					
13					
14					
15					
16					
17					
18					
19					
20					

Practice set 1

Multiple-choice questions

Solutions start on page 152.

Question 1

Six coins are tossed together and the number of heads is noted.

What is the probability that exactly 4 heads appear in one toss?

A $\frac{1}{64}$ **B** $\frac{1}{16}$ **C** $\frac{15}{64}$ **D** $\frac{15}{16}$

Question 2

A binomial random variable is defined as $X \sim \text{Bin}(10, 0.75)$.

Which of the following best describes the shape of the distribution when graphed?

A Positively skewed
B Negatively skewed
C Symmetrical
D There isn't enough information to tell.

Question 3

A binomial random variable is defined as $X \sim \text{Bin}(30, 0.4)$.

What is the mean and variance for this random variable?

A $E(X) = 18$, $\text{Var}(X) = 7.2$
B $E(X) = 12$, $\text{Var}(X) = 7.2$
C $E(X) = 18$, $\text{Var}(X) = 2.7$
D $E(X) = 12$, $\text{Var}(X) = 2.7$

Question 4

The probability that it rains on a particular day is $\frac{3}{10}$.

Which of the following expressions will give the probability that it will not rain on exactly 4 days in a week?

A $\binom{7}{3}(0.3)^3(0.7)^4$ **B** $\binom{7}{4}(0.3)^4(0.7)^3$ **C** $\binom{10}{6}(0.3)^6(0.7)^4$ **D** $(0.3)^4(0.7)^6$

Question 5

Which binomial distribution below is suitable to be approximated by a normal distribution?

A $X \sim \text{Bin}(10, 0.2)$ **B** $X \sim \text{Bin}(20, 0.4)$ **C** $X \sim \text{Bin}(20, 0.85)$ **D** $X \sim \text{Bin}(30, 0.95)$

Question 6

A random variable X defines a binomial distribution where $E(X) = 15$ and $\text{Var}(X) = 3.75$.

Find the values of n and p.

A $n = 20, p = 0.25$ **B** $n = 20, p = 0.75$ **C** $n = 60, p = 0.25$ **D** $n = 60, p = 0.75$

Question 7

A binomial random variable X is defined such that $E(X) = 3.5$ and $\text{Var}(X) = 2.625$.

Calculate $P(X = 4)$.

A 0.000 302 **B** 0.0146 **C** 0.220 **D** 0.396

Question 8

The probability a particular traffic light is red is p. Casey drives past this traffic light every day to get to work. Let the random variable X be the number of times she stops at the traffic light in a working week (5 days).

Which of the following expressions gives the probability that Casey stops at the traffic light at least twice in a week?

A $1 - 5p^4 + 4p^5$ **B** $1 - p^5$ **C** $1 - 5p^4 + 6p^5$ **D** $5p^4 + 4p^5$

Question 9

A six-sided die is rolled 50 times and the number of times a 4 appears is noted.

What is the mean and standard deviation of the binomial distribution formed?

A $E(X) = 8.3, \sigma = 2.6$ **B** $E(X) = 12.5, \sigma = 3.1$

C $E(X) = 8.3, \sigma = 6.9$ **D** $E(X) = 8.3, \sigma = 0.0527$

Question 10

A bag contains red and yellow marbles. Two marbles are drawn from the bag at random, the first replaced before the second one is drawn.

If the probability of drawing at least one red marble is 0.64, which of the following is the ratio of red to yellow marbles?

A 2:5 **B** 3:5 **C** 2:3 **D** 3:2

Question 11

An archer fires at a target with a 90% success rate. He fires 20 shots.

What is the probability he misses at most two times?

A 0.122 **B** 0.392 **C** 0.677 **D** 0.867

Question 12

The proportion of students at a school who wear glasses is 0.40.

Which of the following is the smallest sample size such that the standard deviation is less than 0.05?

A 5 **B** 20 **C** 96 **D** 97

Question 13

The probability that a person can complete an obstacle course is 0.7. A team of 5 people progresses to the next round if a majority of them complete the obstacle course.

What is the probability the team progresses to the next round?

A 2.1% **B** 13% **C** 27% **D** 84%

Question 14

Let $\hat{p}$ be the sample proportion representing the number of students who have a part-time job. A sample of 100 students is taken and it is found that the standard deviation of this sample proportion is approximately 0.033 63.

Which of the following can be the proportion of students who have a part-time job?

A 0.68 **B** 0.74 **C** 0.87 **D** 0.91

Question 15

Let $\hat{p}$ be the sample proportion representing the number of people who shop at the local supermarket. It is known that 80% of people who live in the town shop at the local supermarket.

In a sample of 100 people who live in the town, find the probability that fewer than 68 people shop at the local supermarket.

A 0.15%

B 2.5%

C 16%

D 50%

Question 16

A bag contains 7 red pens and 3 blue pens. Two pens are drawn from the bag at random, the colour is noted and the pens are then replaced back in the bag. This is performed 10 times.

What is the probability that the pens are different colours at least 2 times?

A 0.006 088

B 0.018 15

C 0.9818

D 0.9939

Question 17

A multiple-choice test consists of two sections. Part A has 5 questions with 4 possible answers. Part B has 10 questions with 5 possible answers. Kelly needs to score 50% or more in both sections to pass this test.

If she chooses all her responses randomly, what is the probability that she passes the exam?

A 0.003 395

B 0.0328

C 0.1035

D 0.4

Question 18

A binomial random variable, X, is defined such that the expected value is twice the standard deviation.

Which of the following statements is true?

A $p = \frac{1}{2}$

B $p = \frac{1}{4n+1}$

C $p = \frac{4}{n+4}$

D $p = \frac{2}{n+2}$

Question 19

A beginner darts player has a 25% chance of hitting the centre of the target. He makes 15 throws at the target.

For which 2 numbers of hits of the target will their probability be the same?

A 3 and 4

B 7 and 8

C 8 and 9

D 11 and 12

Question 20

Given a binomial random variable X such that $E(X) = 5$ and $\text{Var}(X) = 2.5$, which value of x below is such that $P(X = x) = 0.205$?

A 5

B 6

C 7

D 8

Practice set 2

Short-answer questions

Solutions start on page 156.

Question 1 (2 marks)

The probability that it will be sunny on a particular day is 0.7.

In a 7-day week, what is the probability that there are more sunny days than non-sunny days? 2 marks

Question 2 (3 marks) ©NESA 2020 HSC EXAM, QUESTION 12(b)

When a particular biased coin is tossed, the probability of obtaining a head is $\frac{3}{5}$.

This coin is tossed 100 times.

Let X be the random variable representing the number of heads obtained.
This random variable will have a binomial distribution.

a Find the expected value, $E(X)$. 1 mark

b By finding the variance, $\text{Var}(X)$, show that the standard deviation of X is approximately 5. 1 mark

c By using a normal approximation, find the approximate probability that X is between 55 and 65. 1 mark

Question 3 (5 marks)

A fair coin is tossed 30 times and X is the random variable representing the number of heads tossed.

a Calculate the mean and variance for this binomial distribution. 2 marks

b What is the probability of getting a result within 1 standard deviation of the mean, correct to four decimal places? 3 marks

Question 4 (4 marks)

a Find the probability that a tail will appear exactly 5 times when 10 coins are tossed together, correct to three decimal places. 1 mark

b Find how many tosses are required to have a 90% chance of tossing exactly 5 tails at least once. 3 marks

Question 5 (2 marks)

X is a random variable representing a Bernoulli trial where the mean is 3 times the standard deviation.

Calculate the value of p. 2 marks

Question 6 (3 marks)

X is a random variable representing a Bernoulli trial where $\text{Var}(X) = \frac{20}{81}$.

Calculate the value of p if $p < 0.5$. 3 marks

PRACTICE SET 2

Question 7 (4 marks)

It is known that 10% of students in a school are left-handed. A sample of 100 students is taken.

a Find the expected value and standard deviation of this sample proportion. 2 marks

b What is the probability that less than 7% of students are left-handed in a given sample? Use the empirical rule to estimate this sample proportion to the normal distribution. 2 marks

Question 8 (7 marks)

A biased die has a $\frac{4}{5}$ chance of landing on a 1. The die is rolled 100 times. Let X be the random variable representing the number of 1s rolled and $\hat{p}$ be the sample proportion of this binomial distribution.

a Find the expected value and variance for the random variable. 2 marks

b Find the expected value and variance for the sample proportion. 2 marks

c Hence, use the normal approximation for the sample proportion to find the probability that the proportion of 1s rolled is between 72% and 84%. 3 marks

Question 9 (5 marks) ©NESA 2004 HSC EXAM, QUESTION 4(c)

Katie is one of 10 members of a social club. Each week, one member is selected at random to win a prize.

a What is the probability that in the first 7 weeks Katie will win at least 1 prize? 1 mark

b Show that in the first 20 weeks Katie has a greater chance of winning exactly 2 prizes than of winning exactly 1 prize. 2 marks

c For how many weeks must Katie participate in the prize drawing so that she has a greater chance of winning exactly 3 prizes than of winning exactly 2 prizes? 2 marks

Question 10 (5 marks)

It is known that a new vaccine is 80% effective in treating a disease. A medical trial is run with 100 people to check its effectiveness.

a What is the mean and standard deviation of the sample proportion of this medical trial? 2 marks

b Use the normal distribution table below to find the probability that it is effective for more than 90% of the trialled population. 3 marks

First decimal place

z	0.0	0.1	0.2	0.3	0.4	0.5	0.6	0.7	0.8	0.9
0	0.5000	0.5398	0.5793	0.6179	0.6554	0.6915	0.7257	0.7580	0.7881	0.8159
1	0.8413	0.8643	0.8849	0.9032	0.9192	0.9332	0.9452	0.9554	0.9641	0.9713
2	0.9772	0.9821	0.9861	0.9893	0.9918	0.9938	0.9953	0.9965	0.9974	0.9981
3	0.9987	0.9990	0.9993	0.9995	0.9997	0.9998	0.9998	0.9999	0.9999	1.0000

Question 11 (4 marks)

The current unemployment rate in Sydney is 6%.

If a random sample of 150 Sydney residents is taken, use a normal approximation to the sample proportion to find the probability that more than 15 are unemployed. Use the table of probabilities below. 4 marks

x	0.00	0.01	0.02	0.03	0.04	0.05	0.06	0.07	0.08	0.09
2.0	0.9772	0.9778	0.9783	0.9788	0.9793	0.9798	0.9803	0.9808	0.9812	0.9817
2.1	0.9821	0.9826	0.9830	0.9834	0.9838	0.9842	0.9846	0.9850	0.9854	0.9857
2.2	0.9861	0.9864	0.9868	0.9871	0.9875	0.9878	0.9881	0.9884	0.9887	0.9890
2.3	0.9893	0.9896	0.9898	0.9901	0.9904	0.9906	0.9909	0.9911	0.9913	0.9916
2.4	0.9918	0.9920	0.9922	0.9925	0.9927	0.9929	0.9931	0.9932	0.9934	0.9936

Question 12 (5 marks)

A phone company takes a sample of 250 households to determine how many use their services. It is known that 20% of all households use their services.

a What are the mean and standard deviation of the sample proportion? 2 marks

b What is the probability that the proportion of households using their services in a given sample is within 3% of the expected value? Use the table of probabilities below. 3 marks

First decimal place

z	0.0	0.1	0.2	0.3	0.4	0. 5	0.6	0.7	0.8	0.9
0	0.5000	0.5398	0.5793	0.6179	0.6554	0.6915	0.7257	0.7580	0.7881	0.8159
1	0.8413	0.8643	0.8849	0.9032	0.9192	0.9332	0.9452	0.9554	0.9641	0.9713
2	0.9772	0.9821	0.9861	0.9893	0.9918	0.9938	0.9953	0.9965	0.9974	0.9981
3	0.9987	0.9990	0.9993	0.9995	0.9997	0.9998	0.9998	0.9999	0.9999	1.0000

Question 13 (4 marks)

It is known that 25% of cars from a certain manufacturer will develop a defect in the first year of use.

Using the normal approximation to the sample proportion, find the probability that, of 100 cars, at most 35 will develop a defect in the first year. Use the table of probabilities below. 4 marks

x	0.00	0.01	0.02	0.03	0.04	0.05	0.06	0.07	0.08	0.09
2.0	0.9772	0.9778	0.9783	0.9788	0.9793	0.9798	0.9803	0.9808	0.9812	0.9817
2.1	0.9821	0.9826	0.9830	0.9834	0.9838	0.9842	0.9846	0.9850	0.9854	0.9857
2.2	0.9861	0.9864	0.9868	0.9871	0.9875	0.9878	0.9881	0.9884	0.9887	0.9890
2.3	0.9893	0.9896	0.9898	0.9901	0.9904	0.9906	0.9909	0.9911	0.9913	0.9916
2.4	0.9918	0.9920	0.9922	0.9925	0.9927	0.9929	0.9931	0.9932	0.9934	0.9936

Question 14 (3 marks)

A machine that packages 1 kg bags of flour is calibrated to a mean of 1.100 kg and a standard deviation of 0.050 kg. A bag is deemed acceptable if the weight is greater than or equal to 1 kg. A sample of 10 bags is taken.

What is the probability that at least 8 bags are of acceptable weight? 3 marks

Question 15 (3 marks)

A survey of 2500 Australian voters is conducted to determine the probability of voting for Party A in the election. It is estimated 20% of the population favours Party A.

Use the normal approximation to the sample proportion to find the probability that the result is within 2% of the estimated amount. Note: $P(z < 2.5) = 0.9938$. 3 marks

Question 16 (6 marks) ©NESA 2006 HSC EXAM, QUESTION 6(b)

In an endurance event, the probability that a competitor will complete the course is p and the probability that a competitor will not complete the course is $q = 1 - p$. Teams consist of either two or four competitors. A team scores points if at least half of its members complete the course.

a Show that the probability that a four-member team will have at least three of its members not complete the course is $4pq^3 + q^4$. 1 mark

b Hence, or otherwise, find an expression in terms of q only for the probability that a four-member team will score points. 2 marks

c Find an expression in terms of q only for the probability that a two-member team will score points. 1 mark

d Hence, or otherwise, find the range of values of q for which a two-member team is more likely than a four-member team to score points. 2 marks

Question 17 (3 marks)

Two identical jars contain blue and red marbles. Jar A contains 4 blue marbles and 3 red marbles. Jar B contains 2 blue and 2 red marbles. Santi chooses a jar at random, then picks one marble out of the jar at random. Santi is twice as likely to choose Jar A than Jar B.

a Find the probability that Santi picks a red marble. 1 mark

b Santi performs this experiment 100 times. What is the probability that she has picked a red marble exactly 50 times? Leave your answer in exact form. 2 marks

Question 18 (5 marks)

A speed camera placed in a 40 km/h zone detects speeds and automatically issues fines when cars are travelling 2 km/h over the speed limit. It is known that cars travel along this road at speeds that follow a normal distribution with a mean of 38 and a standard deviation of 2.

a What percentage of cars will be issued with a fine? 2 marks

b A glitch in the speed camera causes it to malfunction 30% of the time and fines are issued incorrectly. When the camera glitches, cars travelling over the speed limit are not issued a fine, whereas cars travelling at or below the speed limit are issued a fine. If 10 cars pass the speed camera, what is the probability that at least one car is fined? 3 marks

Question 19 (5 marks)

Kelly catches and tags 30 fish, then releases them back into the lake. Two days later, she catches 15 fish in the lake and finds that 2 of them are tagged.

a Determine the population of fish in the lake. 1 mark

b Show that the probability that at most 3 fish are tagged in a catch of 15 fish is 0.871. 2 marks

c Kelly continues to catch groups of 15 fish to check the number tagged. If she made 20 catches, what is the probability that exactly 4 catches yielded more than 3 tagged fish? 2 marks

Question 20 (6 marks)

A factory produces lightbulbs that have a mean lifetime of 1000 hours and a standard deviation of 50 hours. Lightbulbs are said to be faulty if their lifetime is less than 950 hours.

Use the table of probabilities below to answer the following questions.

x	0.00	0.01	0.02	0.03	0.04	0.05	0.06	0.07	0.08	0.09
1.0	0.8413	0.8438	0.8461	0.8484	0.8508	0.8531	0.8554	0.8577	0.8599	0.8621
1.1	0.8643	0.8665	0.8686	0.8708	0.8729	0.8749	0.8770	0.8790	0.8810	0.8830
1.2	0.8849	0.8869	0.8888	0.8907	0.8925	0.8944	0.8962	0.8980	0.8997	0.9015
1.3	0.9032	0.9049	0.9066	0.9082	0.9099	0.9115	0.9131	0.9147	0.9162	0.9177
1.4	0.9192	0.9207	0.9222	0.9236	0.9251	0.9265	0.9279	0.9292	0.9306	0.9319

a Find the probability that a randomly chosen light bulb will be faulty. 2 marks

b A sample of 1500 lightbulbs is taken and the number of faulty lightbulbs is checked. 2 marks

Calculate correct to 3 significant figures the mean and standard deviation for this sample proportion.

c Using the normal approximation to the sample proportion, determine the probability that at most 220 light bulbs from the sample in part **b** are faulty. 2 marks

PRACTICE SET 2

Practice set 1

Worked solutions

1 C

$$P(4 \text{ heads}) = \binom{6}{4}\left(\frac{1}{2}\right)^4\left(\frac{1}{2}\right)^2 = \frac{15}{64}$$

2 B

Since $p > \frac{1}{2}$ and $n = 10$, the distribution will be skewed to the right, which means it is negatively skewed.

3 B

$$E(X) = np = 30 \times 0.4 = 12$$

$$\text{Var}(X) = npq = 30 \times 0.4 \times 0.6 = 7.2$$

4 A

$$P(\text{no rain on 4 days}) = P(\text{rains on 3 days}) = \binom{7}{3}(0.3)^3(0.7)^4$$

5 B

A: $np = 10 \times 0.2 = 2$

$nq = 10 \times 0.8 = 8$

B: $np = 20 \times 0.4 = 8$

$nq = 20 \times 0.6 = 12$

C: $np = 20 \times 0.85 = 17$

$nq = 20 \times 0.15 = 3$

D: $np = 30 \times 0.95 = 28.5$

$nq = 30 \times 0.05 = 1.5$

Binomial distribution can be approximated to a normal distribution if $np \geq 5$ and $nq \geq 5$. Therefore the answer is B.

6 B

$E(X) = 15 = np$ [1]

$\text{Var}(X) = 3.75 = npq$ [2]

$[2] \div [1]$: $q = \frac{3.75}{15} = 0.25$

$p = 1 - 0.25 = 0.75$

Substitute p into [1]:

$0.75n = 15$

$n = 15 \div 0.75 = 20$

7 C

$E(X) = 3.5 = np$ [1]

$\text{Var}(X) = 2.625 = npq$ [2]

$[2] \div [1]$: $q = \frac{2.625}{3.5} = 0.75$

$p = 1 - 0.75 = 0.25$

Substitute p into [1]:

$0.25n = 3.5$

$n = 14$

$$P(X = 4) = \binom{14}{4}(0.25)^4(0.75)^{10} = 0.220$$

8 A

$$\begin{aligned} P(X \geq 2) &= 1 - P(X = 0) - P(X = 1) \\ &= 1 - \binom{5}{0}p^5 - \binom{5}{1}p^4(1 - p) \\ &= 1 - p^5 - 5p^4(1 - p) \\ &= 1 - p^5 - 5p^4 + 5p^5 \\ &= 1 - 5p^4 + 4p^5 \end{aligned}$$

9 A

$$p = \frac{1}{6},\ n = 50$$

$$\begin{aligned} E(x) &= np \\ &= 50 \times \frac{1}{6} \\ &= 8.333 \\ &\approx 8.3 \end{aligned}$$

$$\begin{aligned} \text{Var}(X) &= npq \\ &= 50 \times \frac{1}{6} \times \frac{5}{6} \\ &= 6.944\ldots \end{aligned}$$

$$\begin{aligned} \sigma &= \sqrt{6.9444\ldots} \\ &= 2.64\ldots \\ &\approx 2.6 \end{aligned}$$

10 C

Let p be the probability of drawing a red marble.

$$\begin{aligned} P(\text{at least 1 red}) &= 1 - P(\text{no red}) \\ &= 1 - (1-p)^2 \\ 1 - (1-p)^2 &= 0.64 \\ (1-p)^2 &= 0.36 \\ 1 - p &= \pm 0.6 \\ p &= 1 \pm 0.6 \\ &= 0.4, 1.6 \end{aligned}$$

but $0 < p < 1$.
So $p = 0.4$.

$$\begin{aligned} \text{Red} : \text{Yellow} &= 0.4 : 0.6 \\ &= 2 : 3 \end{aligned}$$

11 C

$$\begin{aligned} P(\leq 2 \text{ misses}) &= P(0 \text{ misses}) + P(1 \text{ miss}) + P(2 \text{ misses}) \\ &= \binom{20}{0} 0.9^{20} 0.1^{0} + \binom{20}{1} 0.9^{19} 0.1^{1} + \binom{20}{2} 0.9^{18} 0.1^{2} \\ &= 0.6769\ldots \\ &\approx 0.677 \end{aligned}$$

Hint
In this solution, we are looking at misses so we define $p = 1 - 0.9 = 0.1$.

12 D

$$\begin{aligned} \sigma &< 0.05 \\ \sqrt{\frac{pq}{n}} &< 0.05 \\ \frac{0.4 \times 0.6}{n} &< 0.0025 \\ 0.0025n &> 0.24 \\ n &> 96 \end{aligned}$$

So $n = 97$.

13 D

$$\begin{aligned} & P(\text{team progresses}) \\ &= P(3 \text{ finish}) + P(4 \text{ finish}) + P(5 \text{ finish}) \\ &= \binom{5}{3}(0.7)^3(0.3)^2 + \binom{5}{4}(0.7)^4(0.3) + \binom{5}{5}(0.7)^5 \\ &= 0.8369 \\ &\approx 84\% \end{aligned}$$

14 C

$\sigma = 0.033\,63$

$$\begin{aligned} \sqrt{\frac{pq}{n}} &= 0.03363 \\ \frac{pq}{100} &= 0.001131 \\ pq &= 0.1131 \\ p(1-p) &= 0.1131 \\ p - p^2 &= 0.1131 \\ p^2 - p + 0.1131 &= 0 \\ p &= \frac{1 \pm \sqrt{1 - 4(0.1131)}}{2} \\ &= 0.13, 0.87 \end{aligned}$$

15 A

$n = 100, p = 0.8$

$E(\hat{p}) = 0.8$

$$\text{Var}(\hat{p}) = \frac{0.8 \times 0.2}{100}$$
$$= 0.0016$$
$$\sigma = 0.04$$

We want to find $P(\hat{p} < 0.68)$:

$$z = \frac{0.68 - 0.8}{0.04}$$
$$= -3$$

$$P(z < -3) = \frac{100\% - 99.7\%}{2} = 0.15\%$$

(using the empirical values on the HSC exam reference sheet)

16 C

$$P(\text{pens the same colour}) = \frac{7}{10} \times \frac{6}{9} + \frac{3}{10} \times \frac{2}{9}$$
$$= \frac{8}{15}$$

$$P(\text{pens different colour}) = 1 - \frac{8}{15}$$
$$= \frac{7}{15}$$

P(different colours at least twice)

$= 1 - P(\text{same each time}) - P(\text{different once})$

$$= 1 - {}^{10}C_0\left(\frac{8}{15}\right)^{10} - {}^{10}C_1\left(\frac{8}{15}\right)^9\left(\frac{7}{15}\right)$$
$$= 0.9818$$

17 A

Part A:

$$P(\text{pass}) = \binom{5}{3}(0.25)^3(0.75)^2 + \binom{5}{4}(0.25)^4(0.75) + \binom{5}{5}(0.25)^5$$
$$= 0.1035$$

Part B:

$$P(\text{pass}) = \binom{10}{5}(0.2)^5(0.8)^5 + \binom{10}{6}(0.2)^6(0.8)^4 + \binom{10}{7}(0.2)^7(0.8)^3 + \binom{10}{8}(0.2)^8(0.8)^2 + \binom{10}{9}(0.2)^9(0.8) + \binom{10}{10}(0.2)^{10}$$
$$= 0.0328$$

$$P(\text{pass the exam}) = 0.1035 \times 0.0328$$
$$= 0.003\,395$$

18 C

$E(X) = 2\sigma$

$$np = 2\sqrt{npq}$$
$$n^2p^2 = 4npq$$
$$np = 4q$$ (since $n \neq 0$ and $p \neq 0$ as either value will give a trivial solution)
$$np = 4(1 - p)$$
$$np = 4 - 4p$$
$$p(n + 4) = 4$$

So $p = \dfrac{4}{n + 4}$.

19 A

$$P(\text{hits the target } r \text{ times}) = \binom{15}{r}\left(\frac{1}{4}\right)^r\left(\frac{3}{4}\right)^{n-r}$$

Hint

An alternative method is to test each answer A, B, C and D.

$$P(\text{hits the target } r+1 \text{ times}) = \binom{15}{r+1}\left(\frac{1}{4}\right)^{r+1}\left(\frac{3}{4}\right)^{n-r-1}$$

$$\binom{15}{r}\left(\frac{1}{4}\right)^r\left(\frac{3}{4}\right)^{n-r} = \binom{15}{r+1}\left(\frac{1}{4}\right)^{r+1}\left(\frac{3}{4}\right)^{n-r-1}$$

$$\frac{15!}{r!(15-r)!}\times\frac{1}{4^r}\times\frac{3^{n-r}}{4^{n-r}} = \frac{15!}{(r+1)!(15-r-1)!}\times\frac{1}{4^{r+1}}\times\frac{3^{n-r-1}}{4^{n-r-1}}$$

$$\frac{3^{n-r}}{4^n(15-r)} = \frac{3^{n-r-1}}{4^n(r+1)}$$

$$\frac{3}{15-r} = \frac{1}{r+1}$$

$$3(r+1) = 15-r$$

$$3r+3 = 15-r$$

$$4r = 12$$

$$r = 3$$

$$r+1 = 4$$

So probabilities are equal for 3 and 4 hits of the target in 15 throws.

20 B

$E(X) = 5 = np$ [1]

$\text{Var}(X) = 2.5 = npq$ [2]

[2] ÷ [1]: $q = 0.5$

then $p = 0.5$.

Substitute into [1]: $0.5p = 5$

$p = 10$.

$P(X = x) = 0.205$

$$\binom{10}{x}(0.5)^x(0.5)^{10-x} = 0.205$$

$$\binom{10}{x}(0.5)^{10} = 0.205$$

$$\binom{10}{x} = 209.92$$

$$\approx 210$$

So $x \approx 4$ or 6.

Practice set 2

Worked solutions

Question 1

$P(X \geq 4)$

$= \binom{7}{4}0.7^4 \times 0.3^3 + \binom{7}{5}0.7^5 \times 0.3^2 + \binom{7}{6}0.7^6 \times 0.3^1 + \binom{7}{7}0.7^7 \times 0.3^0$

$= 0.874$

Question 2

a $E(X) = np$

$= 100 \times \frac{3}{5}$

$= 60$

b $\text{Var}(X) = npq$

$= 100 \times \frac{3}{5} \times \frac{2}{5}$

$= 24$

$\sigma = \sqrt{24}$

$= 4.899$

≈ 5

c $P(55 < X < 65)$

$= P\left(\frac{55-60}{5} < Z < \frac{65-60}{5}\right)$

$= P(-1 < Z < 1)$

$= 68\%$ using the empirical values

Question 3

a $E(X) = np$

$= 30 \times \frac{1}{2}$

$= 15$

$\text{Var}(X) = npq$

$= 30 \times \frac{1}{2} \times \frac{1}{2}$

$= 7.5$

b $\sigma = \sqrt{7.5}$

≈ 2.74

To be within 1 standard deviation is 15 ± 2.74, so between 12.26 and 17.74.

The discrete values between 12.26 and 17.74 are 13, 14, 15, 16 and 17.

$P(\text{within 1 standard deviation})$

$= P(X \geq 13)$

$= \binom{30}{13}(0.5)^{13}(0.5)^{17} + \binom{30}{14}(0.5)^{14}(0.5)^{16} + \binom{30}{15}(0.5)^{15}(0.5)^{15} + \binom{30}{16}(0.5)^{16}(0.5)^{14} + \binom{30}{17}(0.5)^{17}(0.5)^{13}$

≈ 0.6384

Question 4

a $P(X = 5)$

$= \binom{10}{5}(0.5)^5(0.5)^5$

$= 0.246$

b $P(X \neq 5)$

$= 1 - 0.246$

$= 0.754$

$P(5 \text{ tails at least once})$

$= 1 - P(X \neq 5)$

$= 1 - (0.754)^n$

$$\begin{aligned} P(5 \text{ tails at least once}) &\geq 0.9 \\ 1 - (0.754)^n &\geq 0.9 \\ (0.754)^n &\leq 0.1 \\ n \ln 0.754 &\leq \ln 0.1 \\ n &\geq \frac{\ln 0.1}{\ln 0.754} \quad (\ln 0.754 < 0) \\ n &\geq 8.15 \end{aligned}$$

9 tosses are needed to have a 90% chance of tossing exactly 5 tails at least once.

Question 5

$E(X) = p, \sigma = \sqrt{pq}$

$$\begin{aligned} p &= 3\sqrt{pq} \\ p^2 &= 9p(1-p) \\ p^2 &= 9p - 9p^2 \\ 10p^2 - 9p &= 0 \\ p(10p - 9) &= 0 \end{aligned}$$

$p = 0, p = \frac{9}{10}$

Since $0 < p < 1, p = \frac{9}{10}$

Question 6

$\text{Var}(X) = \frac{20}{81} = pq$

$$\begin{aligned} p(1-p) &= \frac{20}{81} \\ 81p - 81p^2 &= 20 \\ 81p^2 - 81p + 20 &= 0 \\ p &= \frac{81 \pm \sqrt{81^2 - 4 \times 81 \times 20}}{2 \times 81} \\ &= \frac{4}{9}, \frac{5}{9} \end{aligned}$$

Since $p < 0.5, p = \frac{4}{9}$.

Question 7

a $E(\hat{p}) = 0.1$

$\sigma = \sqrt{\frac{0.1 \times 0.9}{100}}$

$= 0.03$

b Find $P(\hat{p} < 0.07)$:

$z = \frac{0.07 - 0.1}{0.03}$

$= -1$

$P(z < -1) = \frac{100\% - 68\%}{2}$

$= 16\%$

Question 8

a $E(X) = np$

$= 100 \times \frac{4}{5}$

$= 80$

$\text{Var}(X) = npq$

$= 100 \times \frac{4}{5} \times \frac{1}{5}$

$= 16$

b $E(\hat{p}) = p = \frac{4}{5}$

$\text{Var}(\hat{p}) = \frac{pq}{n}$

$= \frac{\frac{4}{5} \times \frac{1}{5}}{100}$

$= 0.0016$

c Find z-scores for 72% and 84% with mean $= \frac{4}{5} = 0.8$:

$\sigma = \sqrt{0.0016}$

$= 0.04$

$z_1 = \frac{0.72 - 0.8}{0.04}$

$= -2$

$z_2 = \frac{0.84 - 0.8}{0.04}$

$= 1$

$P(0.72 < \hat{p} < 0.84)$

$= P(-2 < z < 1)$

$= 68\% + 13.5\%$

$= 81.5\%$

WORKED SOLUTIONS

Question 9

a $P(\text{wins at least one prize})$
$= 1 - P(\text{wins no prizes})$
$= 1 - (0.9)^7$
$= 0.5217$

b In first 20 weeks:

$$P(\text{wins one prize}) = \binom{20}{1}(0.1)(0.9)^{19} = 0.270$$

$$P(\text{wins two prizes}) = \binom{20}{2}(0.1)^2(0.9)^{18} = 0.285$$

So $P(\text{wins two prizes}) > P(\text{wins one prize})$.

c $P(\text{wins 3 prizes}) = \binom{n}{3}(0.1)^3(0.9)^{n-3}$

$P(\text{wins 2 prizes}) = \binom{n}{2}(0.1)^2(0.9)^{n-2}$

$$\binom{n}{3}(0.1)^3(0.9)^{n-3} > \binom{n}{2}(0.1)^2(0.9)^{n-2}$$

$$\frac{n!}{(n-3)!3!} \times 0.1 > \frac{n!}{(n-2)!2!} \times 0.9$$

$$\frac{2 \times (n-2)!}{6 \times (n-3)!} > 9$$

$$n - 2 > 27$$

$$n > 29$$

Katie must participate for 30 or more weeks.

Question 10

a $E(\hat{p}) = 0.8$

$$\text{Var}(\hat{p}) = \frac{0.8 \times 0.2}{100} = 0.0016$$

$$\sigma = \sqrt{0.0016} = 0.04$$

b $z = \dfrac{0.9 - 0.8}{0.04} = 2.5$

$P(\hat{p} > 0.9) = P(z > 2.5)$
$= 1 - 0.9938$
$= 0.0062$
$= 0.62\%$

Question 11

$E(\hat{p}) = 0.06$

$$\text{Var}(\hat{p}) = \frac{0.06 \times 0.94}{150} = 0.000\,376$$

$\sigma = 0.0194$

$$\hat{p} = \frac{15}{150} = 0.1$$

$$z = \frac{0.1 - 0.06}{0.0194} = 2.06$$

$P(\hat{p} > 0.15) = P(z > 2.06)$
$= 1 - P(z < 2.06)$
$= 1 - 0.9803$
$= 0.0197$
$= 1.97\%$

Question 12

a $E(\hat{p}) = p = 0.2$

$$\text{Var}(\hat{p}) = \frac{pq}{n} = \frac{0.2 \times 0.8}{250} = 0.000\,64$$

$$\sigma = \sqrt{0.000\,64} = 0.0253$$

b Find $P(0.17 < \hat{p} < 0.23)$:

$$z_1 = \frac{0.17 - 0.2}{0.0253} = -1.2$$

$$z_2 = \frac{0.23 - 0.2}{0.0253} = 1.2$$

$P(0.17 < \hat{p} < 0.23)$
$= P(-1.2 < z < 1.2)$
$= 2P(0 < z < 1.2)$
$= 2(0.8849 - 0.5)$
$= 0.7698$

9780170459242

Question 13

$E(\hat{p}) = 0.25$

$$\text{Var}(\hat{p}) = \frac{0.25 \times 0.75}{100} = 0.001\,875$$

$$\sigma = \sqrt{0.001\,875} = 0.0433$$

Find $P(\hat{p} < 0.35)$:

$$z = \frac{0.35 - 0.25}{0.0433} = 2.31$$

$$P(\hat{p} < 0.35) = P(z < 2.31) = 0.9896$$

Question 14

$\mu = 1.100, \sigma = 0.050$

Find $P(x > 1)$:

$$z = \frac{1 - 1.100}{0.050} = -2$$

$$P(x > 1) = P(x > -2) = 95\% + \frac{100\% - 95\%}{2} = 97.5\% = 0.975$$

$P(\text{bag of acceptable weight}) = 0.975$

$P(\text{at least 8 bags acceptable})$

$= P(X = 8) + P(X = 9) + P(X = 10)$

$$= \binom{10}{8}(0.975)^8(0.025)^2 + \binom{10}{9}(0.975)^9(0.025)^1 + \binom{10}{10}(0.975)^{10}$$

$= 0.9984$

Question 15

$E(\hat{p}) = 0.2$

$$\text{Var}(\hat{p}) = \frac{0.2 \times 0.8}{2500} = 0.000\,064$$

$$\sigma = \sqrt{0.000\,064} = 0.008$$

Find $P(0.18 < \hat{p} < 0.22)$:

$$z_1 = \frac{0.18 - 0.2}{0.008} = -2.5$$

$$z_2 = \frac{0.22 - 0.2}{0.008} = 2.5$$

$$P(0.18 < \hat{p} < 0.22) = P(-2.5 < z < 2.5) = 2P(0 < z < 2.5) = 2(0.9938 - 0.5) = 0.9876$$

WORKED SOLUTIONS

Question 16

a $P(\text{at least 3 not complete})$
$= P(\text{3 not complete}) + P(\text{4 not complete})$
$= P(\text{1 complete}) + P(\text{0 complete})$
$= \binom{4}{1}p^1q^3 + \binom{4}{0}p^0q^4$
$= 4pq^3 + q^4$

b $P(\text{score points}) = P(\text{2, 3 or 4 complete})$
$= 1 - P(\text{0 or 1 complete})$
$= 1 - P(\text{3 or 4 don't complete})$
$= 1 - (4pq^3 + q^4)$ (from part **i**)
$= 1 - 4(1 - q)q^3 - q^4$
$= 1 - 4q^3 + 4q^4 - q^4$
$= 1 - 4q^3 + 3q^4$

c $P(\text{two-member team scores points}) = 1 - P(\text{neither completes the course})$
$= 1 - q^2$

d $P(\text{two-member team scores}) > P(\text{four-member team scores})$
$1 - q^2 > 1 - 4q^3 + 3q^4$
$0 > 3q^4 - 4q^3 + q^2$
$0 > q^2(3q^2 - 4q + 1)$
$0 > q^2(3q - 1)(q - 1)$

So $\frac{1}{3} < q < 1$. (using the sketch of the function below)

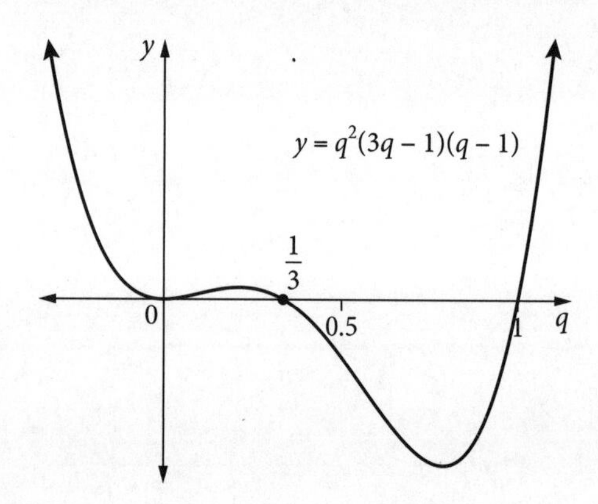

Question 17

a $P(\text{red marble}) = \frac{2}{3} \times \frac{3}{7} + \frac{1}{3} \times \frac{1}{2}$
$= \frac{19}{42}$

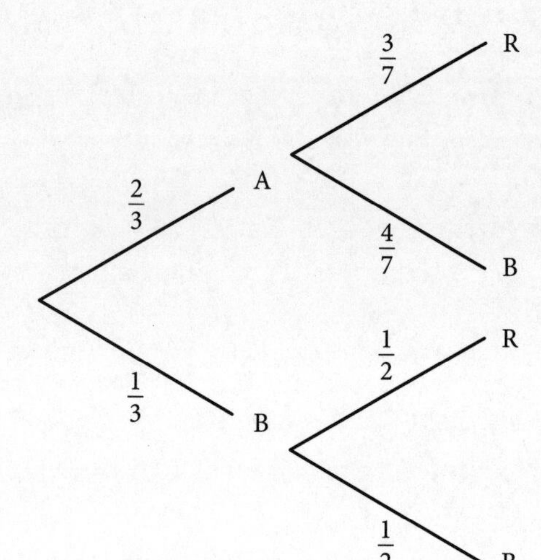

b $P(\text{50 red out of 100}) = \binom{100}{50}\left(\frac{19}{42}\right)^{50}\left(\frac{23}{42}\right)^{50}$

Question 18

a Find $P(x > 42)$:

$$z = \frac{42 - 38}{2}$$
$$= 2$$

$$P(x > 42) = P(z > 2)$$
$$= \frac{100\% - 95\%}{2}$$
$$= 2.5\% \quad \text{(using empirical values)}$$

b

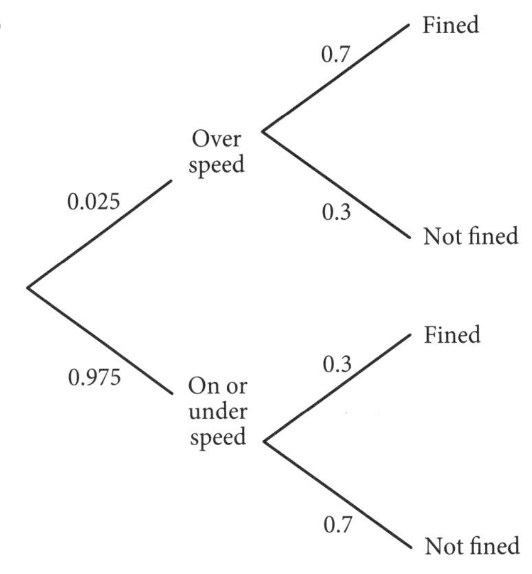

$$P(\text{one car is fined}) = 0.025 \times 0.7 + 0.975 \times 0.3$$
$$= 0.31$$

$$P(\text{at least one car fined out of 10}) = 1 - P(\text{no cars fined})$$
$$= 1 - (1 - 0.31)^{10}$$
$$\approx 0.976$$

Question 19

a $\dfrac{30}{P} = \dfrac{2}{15}$

$$2P = 450$$
$$P = 225$$

b $P(\text{one tagged fish}) = \dfrac{30}{225}$

$$= \frac{2}{15}$$

$$P(\text{at most 3 fish of 15}) = P(x \le 3)$$
$$= \binom{15}{0}\left(\frac{13}{15}\right)^{15} + \binom{15}{1}\left(\frac{13}{15}\right)^{14}\left(\frac{2}{15}\right) + \binom{15}{2}\left(\frac{13}{15}\right)^{13}\left(\frac{2}{15}\right)^{2} + \binom{15}{3}\left(\frac{13}{15}\right)^{12}\left(\frac{2}{15}\right)^{3}$$
$$= 0.8708$$
$$\approx 0.871$$

c $P(\text{4 of 15 catches have more than 3 tagged fish}) = \dbinom{15}{4}(0.871)^{11} \times (1 - 0.871)^{4}$

$$\approx 0.0827$$

Question 20

a $\mu = 1000, \sigma = 50, x = 950$

$$z = \frac{950 - 1000}{50} = -1$$

$$P(z < -1) = 1 - 0.8413$$
$$= 0.1587$$ (or 0.16 if using the empirical values)

b $E(\hat{p}) = p = 0.1587$

$$\text{Var}(\hat{p}) = \frac{pq}{n}$$
$$= \frac{0.1587 \times 0.8413}{1500}$$
$$\approx 0.000089$$
$$\sigma = \sqrt{0.000089}$$
$$\approx 0.00943$$

c Find $P\left(\hat{p} < \frac{220}{1500}\right)$:

$\hat{p} = 0.1467$

$$z = \frac{0.1467 - 0.1587}{0.00943}$$
$$= -1.27$$

$$P\left(\hat{p} < \frac{220}{1500}\right) = P(z < -1.27)$$
$$= 1 - P(z < 1.27)$$
$$= 1 - 0.8980$$
$$= 0.1020$$

HSC exam topic grid (2011–2020)

This table shows the coverage of this topic in past HSC exams by question number. The past exams can be downloaded from the NESA website (www.educationstandards.nsw.edu.au) by selecting 'Year 11 – Year 12', 'HSC exam papers'. NESA marking feedback and guidelines can also be found there.

The new Mathematics Extension 1 course was first examined in 2020. For exams before 2020, select 'Year 11 – Year 12', 'Resources archive', 'HSC exam papers archive'.

Binomial distributions and sample proportion were introduced to the Mathematics Extension 1 course in 2020.

	Binomial probability	Binomial distributions	Sample proportion
2011	6(c)		
2012	12(c)		
2013	11(c)		
2014	11(b)		
2015	14(c)		
2016	11(f)		
2017	11(g)		
2018	12(d)		
2019	11(f)		
2020 new course		**12(b)**	

The question in **bold** can be found on page 147 of this chapter.

HSC exam reference sheet

Mathematics Advanced, Extension 1 and Extension 2

Note: Unlike the actual HSC exam reference sheet, this sheet indicates which formulas are Mathematics Extension 1 and 2.

Measurement

Length

$$l = \frac{\theta}{360} \times 2\pi r$$

Area

$$A = \frac{\theta}{360} \times \pi r^2$$

$$A = \frac{h}{2}(a + b)$$

Surface area

$$A = 2\pi r^2 + 2\pi rh$$

$$A = 4\pi r^2$$

Volume

$$V = \frac{1}{3}Ah$$

$$V = \frac{4}{3}\pi r^3$$

Financial Mathematics

$$A = P(1 + r)^n$$

Sequences and series

$$T_n = a + (n - 1)d$$

$$S_n = \frac{n}{2}\left[2a + (n - 1)d\right] = \frac{n}{2}(a + l)$$

$$T_n = ar^{n-1}$$

$$S_n = \frac{a(1 - r^n)}{1 - r} = \frac{a(r^n - 1)}{r - 1}, r \neq 1$$

$$S = \frac{a}{1 - r}, |r| < 1$$

Functions

$$x = \frac{-b \pm \sqrt{b^2 - 4ac}}{2a}$$

For $ax^3 + bx^2 + cx + d = 0$:* *EXT1

$$\alpha + \beta + \gamma = -\frac{b}{a}$$

$$\alpha\beta + \alpha\gamma + \beta\gamma = \frac{c}{a}$$

$$\text{and } \alpha\beta\gamma = -\frac{d}{a}$$

Relations

$$(x - h)^2 + (y - k)^2 = r^2$$

Logarithmic and Exponential Functions

$$\log_a a^x = x = a^{\log_a x}$$

$$\log_a x = \frac{\log_b x}{\log_b a}$$

$$a^x = e^{x \ln a}$$

Trigonometric Functions

$\sin A = \frac{\text{opp}}{\text{hyp}}, \cos A = \frac{\text{adj}}{\text{hyp}}, \tan A = \frac{\text{opp}}{\text{adj}}$

$A = \frac{1}{2}ab\sin C$

$\frac{a}{\sin A} = \frac{b}{\sin B} = \frac{c}{\sin C}$

$c^2 = a^2 + b^2 - 2ab\cos C$

$\cos C = \frac{a^2 + b^2 - c^2}{2ab}$

$l = r\theta$

$A = \frac{1}{2}r^2\theta$

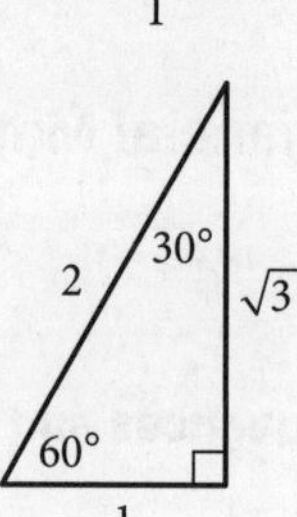

Trigonometric identities

$\sec A = \frac{1}{\cos A}, \cos A \neq 0$

$\operatorname{cosec} A = \frac{1}{\sin A}, \sin A \neq 0$

$\cot A = \frac{\cos A}{\sin A}, \sin A \neq 0$

$\cos^2 x + \sin^2 x = 1$

Compound angles*

$\sin(A + B) = \sin A\cos B + \cos A\sin B$

$\cos(A + B) = \cos A\cos B - \sin A\sin B$

$\tan(A + B) = \frac{\tan A + \tan B}{1 - \tan A\tan B}$

If $t = \tan\frac{A}{2}$, then $\sin A = \frac{2t}{1+t^2}$

$\cos A = \frac{1-t^2}{1+t^2}$

$\tan A = \frac{2t}{1-t^2}$

$\cos A\cos B = \frac{1}{2}[\cos(A - B) + \cos(A + B)]$

$\sin A\sin B = \frac{1}{2}[\cos(A - B) - \cos(A + B)]$

$\sin A\cos B = \frac{1}{2}[\sin(A + B) + \sin(A - B)]$

$\cos A\sin B = \frac{1}{2}[\sin(A + B) - \sin(A - B)]$

$\sin^2 nx = \frac{1}{2}(1 - \cos 2nx)$

$\cos^2 nx = \frac{1}{2}(1 + \cos 2nx)$

Statistical Analysis

$z = \frac{x - \mu}{\sigma}$

An outlier is a score
less than $Q_1 - 1.5 \times IQR$
or
more than $Q_3 + 1.5 \times IQR$

Normal distribution

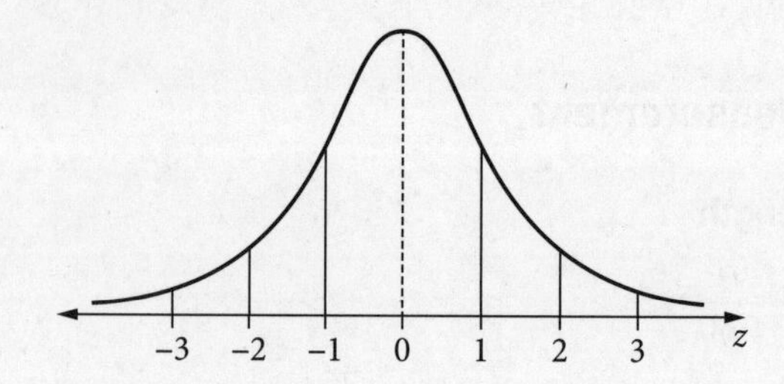

- approximately 68% of scores have z-scores between −1 and 1
- approximately 95% of scores have z-scores between −2 and 2
- approximately 99.7% of scores have z-scores between −3 and 3

Discrete random variables

$E(X) = \mu$

$\operatorname{Var}(X) = E[(X - \mu)^2] = E(X^2) - \mu^2$

Probability

$P(A \cap B) = P(A)P(B)$

$P(A \cup B) = P(A) + P(B) - P(A \cap B)$

$P(A|B) = \frac{P(A \cap B)}{P(B)}, P(B) \neq 0$

Continuous random variables

$P(X \leq r) = \int_a^r f(x)\,dx$

$P(a < X < b) = \int_a^b f(x)\,dx$

Binomial distribution*

$P(X = r) = {}^nC_r p^r(1 - p)^{n-r}$

$X \sim \operatorname{Bin}(n, p)$

$\Rightarrow P(X = x)$

$= \binom{n}{x} p^x(1 - p)^{n-x}, x = 0, 1, \ldots, n$

$E(X) = np$

$\operatorname{Var}(X) = np(1 - p)$

*EXT1

Differential Calculus

Function	Derivative
$y = f(x)^n$	$\frac{dy}{dx} = nf'(x)\left[f(x)\right]^{n-1}$
$y = uv$	$\frac{dy}{dx} = u\frac{dv}{dx} + v\frac{du}{dx}$
$y = g(u)$ where $u = f(x)$	$\frac{dy}{dx} = \frac{dy}{du} \times \frac{du}{dx}$
$y = \frac{u}{v}$	$\frac{dy}{dx} = \frac{v\frac{du}{dx} - u\frac{dv}{dx}}{v^2}$
$y = \sin f(x)$	$\frac{dy}{dx} = f'(x)\cos f(x)$
$y = \cos f(x)$	$\frac{dy}{dx} = -f'(x)\sin f(x)$
$y = \tan f(x)$	$\frac{dy}{dx} = f'(x)\sec^2 f(x)$
$y = e^{f(x)}$	$\frac{dy}{dx} = f'(x)e^{f(x)}$
$y = \ln f(x)$	$\frac{dy}{dx} = \frac{f'(x)}{f(x)}$
$y = a^{f(x)}$	$\frac{dy}{dx} = (\ln a)f'(x)a^{f(x)}$
$y = \log_a f(x)$	$\frac{dy}{dx} = \frac{f'(x)}{(\ln a)f(x)}$
$y = \sin^{-1} f(x)$	$\frac{dy}{dx} = \frac{f'(x)}{\sqrt{1 - [f(x)]^2}}$ *
$y = \cos^{-1} f(x)$	$\frac{dy}{dx} = -\frac{f'(x)}{\sqrt{1 - [f(x)]^2}}$ *
$y = \tan^{-1} f(x)$	$\frac{dy}{dx} = \frac{f'(x)}{1 + [f(x)]^2}$ *

Integral Calculus

$$\int f'(x)\left[f(x)\right]^n dx = \frac{1}{n+1}\left[f(x)\right]^{n+1} + c$$

where $n \neq -1$

$$\int f'(x)\sin f(x)\,dx = -\cos f(x) + c$$

$$\int f'(x)\cos f(x)\,dx = \sin f(x) + c$$

$$\int f'(x)\sec^2 f(x)\,dx = \tan f(x) + c$$

$$\int f'(x)e^{f(x)}\,dx = e^{f(x)} + c$$

$$\int \frac{f'(x)}{f(x)}dx = \ln|f(x)| + c$$

$$\int f'(x)a^{f(x)}\,dx = \frac{a^{f(x)}}{\ln a} + c$$

$$\int \frac{f'(x)}{\sqrt{a^2 - [f(x)]^2}}\,dx = \sin^{-1}\frac{f(x)}{a} + c \; *$$

$$\int \frac{f'(x)}{a^2 + [f(x)]^2}\,dx = \frac{1}{a}\tan^{-1}\frac{f(x)}{a} + c \; *$$

$$\int u\frac{dv}{dx}dx = uv - \int v\frac{du}{dx}\,dx \; **$$

$$\int_a^b f(x)\,dx$$

$$\approx \frac{b-a}{2n}\left\{f(a) + f(b) + 2\left[f(x_1) + \cdots + f(x_{n-1})\right]\right\}$$

where $a = x_0$ and $b = x_n$

*EXT1, **EXT2

9780170459242

Combinatorics*

$${}^nP_r = \frac{n!}{(n-r)!}$$

$$\binom{n}{r} = {}^nC_r = \frac{n!}{r!(n-r)!}$$

$$(x+a)^n = x^n + \binom{n}{1}x^{n-1}a + \cdots + \binom{n}{r}x^{n-r}a^r + \cdots + a^n$$

Vectors*

$$|\underset{\sim}{u}| = \left|x\underset{\sim}{i} + y\underset{\sim}{j}\right| = \sqrt{x^2 + y^2}$$

$$\underset{\sim}{u} \cdot \underset{\sim}{v} = |\underset{\sim}{u}||\underset{\sim}{v}|\cos\theta = x_1x_2 + y_1y_2,$$

where $\underset{\sim}{u} = x_1\underset{\sim}{i} + y_1\underset{\sim}{j}$

and $\underset{\sim}{v} = x_2\underset{\sim}{i} + y_2\underset{\sim}{j}$

$\underset{\sim}{r} = \underset{\sim}{a} + \lambda\underset{\sim}{b}$**

Complex Numbers**

$$z = a + ib = r(\cos\theta + i\sin\theta)$$
$$= re^{i\theta}$$

$$\left[r(\cos\theta + i\sin\theta)\right]^n = r^n(\cos n\theta + i\sin n\theta)$$
$$= r^n e^{in\theta}$$

Mechanics**

$$\frac{d^2x}{dt^2} = \frac{dv}{dt} = v\frac{dv}{dx} = \frac{d}{dx}\left(\frac{1}{2}v^2\right)$$

$$x = a\cos(nt + \alpha) + c$$

$$x = a\sin(nt + \alpha) + c$$

$$\ddot{x} = -n^2(x - c)$$

*EXT1, **EXT2

Index

A

absolute value 24
acceleration 21, 27
acceleration equations 27
areas
  about the x-axis 83
  about the y-axis 84
  bounded by a curve 84–85
associative law (vectors) 25
assumption 4
auxiliary angle formulas 56
auxiliary angle method 54, 77
axis of rotation 84

B

Bernoulli distribution 139, 140–141
Bernoulli random variable 139, 140
Bernoulli trial 139, 140, 142
binomial distribution formulas 164
binomial distributions 139, 140–142, 162
  mean 141
  probabilities 140–141
  skewed, negatively 141
  skewed, positively 141
  symmetrical 141
binomial probability 162
binomial random variable 139, 140, 142

C

Cartesian equation 21
  of the path 28–29
column vector notation 21
column vectors 24
commutative law (vectors) 23, 25
component form 21
compound angle 55
compound angle identities 54, 55, 164
constant solution 110

D

derivative 110, 165
differential equations 109, 110–112
  applications of 110–112, 136
  second-order 111
  separation of variables 109
  solving 136
  solving $\frac{dy}{dx} = f(x)$ 110
  solving $\frac{dy}{dx} = f(x)g(y)$ 110–111
  solving $\frac{dy}{dx} = g(y)$ 110
direction field 109, 111, 136
discrete probability distribution 142
discrete random variable 140, 142
displacement 21
displacement vectors 21, 22
distributive law (vectors) 25
divisibility 3, 4
domain 54, 56
dot product
  See scalar product
double angle 55
double angle identities 54, 55

E

exponential function 109
exponential growth and decay 109, 112, 136

F

first derivative 111
first-order differential equations 111

G

general solution of a trigonometric equation 56
gravity 21

I

identities 54
  proving 77
  trigonometric 164
initial displacement 28
initial position 21
initial speed 27
initial velocity 21
integrand 80
integration
  by substitution 80, 81, 106
  of $\sin^2 x$ and $\cos^2 x$ 82
inverse functions 80, 83, 106
inverse trigonometric functions 83, 97, 106
  derivatives 83
  integrals 83

L

limits 84
logistic equations 109, 112

M

magnitude
  of a vector 21, 24
mathematical induction 3, 4
maximum height 21, 27
mean 139, 141

N

Newton's law of cooling 109, 112
normal distribution 142
normal to a curve 80

O

orthogonal 21
orthogonal projection 27

P

parallelogram law
  of addition 23
  of subtraction 23
parallel vectors 21, 22
particle 21, 29
path 21
perpendicular vectors 26
population proportion 142
position vector 21, 24
  of a curve 29
probability distribution
  discrete 139, 140
product of powers
  sine and cosine ratios 81
  tangent and secant ratios 81
products to sums identities 54, 56, 82
projectile 21
projectile motion 27–29, 51
  equations of 28
  vector functions for 29
projection of a vector 26–27
proof by mathematical induction 4–6
proofs
  divisibility 5–6, 20
  false 4
  geometric 26
  series 4–5, 20

R

range 21
reciprocal 82, 110
reverse chain rule 80, 83
RTP 3

S

sample proportions 139, 142
  normal approximation 139, 142
sampling distribution of proportions 142
  mean 142
  variance 142
scalar multiplication 23, 25
scalar product 21, 25
  properties of 25
scalar projection 21, 27
scalars 21, 22
second derivative 111
slope field
  See direction field
solid of revolution 80
standard deviation 139
statement 3, 4
subtracting vectors 23–24

T

tangent to a curve 80
t-formulas 54, 56
time of flight 21
transformation 56
triangle law
  of addition (vectors) 23
  of subtraction (vectors) 23
trigonometric
  derivatives 82
  integrals 82, 106
trigonometric equations 55–56
  general solution of 56
  using identities 77
trigonometric functions
  inverse 83

U

unit vectors 21, 24

V

variance 139, 142
vectors 21, 22
  adding 23–24
  angle between 2 vectors 26
  applying 51
  displacement 22
  equal 21, 22
  length and angle 25
  operations with 51
  parallel 22
  perpendicular 26
  projection of 21, 26–27
  standard unit 26
  subtracting 23–24
velocity 21, 27
volumes
  rotated about the x-axis 85
  rotated about the y-axis 85
volumes of solids of revolution 83–85, 106

W

wave function 56

Z

zero vector 23
z-scores 142